DO YOU

2. IM
3. SATELLITE COMPUTER
4. AUTOSIZING
5. KLUDGE
6. TARGET DISK

1. Allowing the same image to remain on the screen for long periods of time, so that a shadow of the image is etched permanently into the screen's phosphorous coating. Burn-in can be distracting to the user, and will interfere with the clarity of screen images.

2. The process of creating, storing, and manipulating images, particularly exact duplicates of images such as legal forms, photographs, or X rays.

3. Another way of referring to a computer at a remote location that is attached to the main computer.

4. In computer graphics, the automatic adjustment of images to fit the size of the computer screen.

5. A computer system difficult to use because of poor planning and design.

6. The disk on which data are copied during the process of copying data from one disk to another. The disk the data are copied from is the source disk.

WITH COMPUTERS IN EVERY OFFICE, SCHOOL, AND WORKPLACE AS WELL AS MANY HOMES, THE *21ST CENTURY DIC-TIONARY OF COMPUTER TERMS* GIVES YOU EASY ACCESS TO THE SPECIAL LAN-GUAGE OF TODAY'S TECHNOLOGY.

21ST CENTURY DICTIONARY
OF COMPUTER TERMS

—21st— CENTURY DICTIONARY OF COMPUTER TERMS

EDITED BY THE PRINCETON LANGUAGE INSTITUTE

GARY McCLAIN, Ph.D., COMPILER

Produced by The Philip Lief Group, Inc.

A LAUREL BOOK

Published by
Dell Publishing
a division of
Bantam Doubleday Dell Publishing Group, Inc.
1540 Broadway
New York, NY 10036

Published by arrangement with
The Philip Lief Group, Inc.
6 West 20th Street
New York, NY 10011

ISBN: 0-440-21557-9

Printed in the United States of America

Published simultaneously in Canada

June 1994

10 9 8 7 6 5 4 3 2 1

Contents

Contents

Introduction

Computers, the foundation of the coming "Information Superhighway," have a vocabulary of their own. New words, acronyms, and applications evolve at such a fast pace that users, of all levels and experience, find themselves searching trade publications or asking colleagues for definitions. For the beginning or occasional computer user, vocabulary can be even more daunting.

Further complicating matters, each area of high technology has its own separate vocabulary. For example, those working at the mainframe, or enterprise, level of computing have one set of terms and concepts. Personal computers have another—and within it, DOS and Macintosh also have their own distinct subsets of terminology. Application areas such as word processing, graphics, and networking offer further vocabulary challenges.

The goal of the *21st Century Dictionary of Computer Terms* is to demystify computer terminology. The most commonly used computer terms in all areas and disciplines are clearly explained with understandable, non-jargony, concise definitions. For ease of use, acronyms, and the full terms they refer to, are both included separately in the dictionary's alphabetical listings. Additionally, synonymous terms are cross-referenced. Those terms that are unique to a specific category of computing are grouped together and labeled as such in the "Index of Computer Terms." These categories, and their labels, are as follows:

Database and Spreadsheet;

Graphics and Video;

Large Systems, including mainframe and midrange computers, and concepts related to computing in a large organization environment;

Networking and Communications, at any hardware level, from local area networks to multiple mainframes;

Operating Systems, including DOS and UNIX;

Personal Computers, including hardware specifically related to personal computers, as well as laptop, notebook, and other portable computers;

Printers and Peripheral Devices, including fax machines and modems, and the hardware and software terms related to connecting and using these peripheral devices;

Programming, including any terms related to computer programming/software application development; and

Word Processing and Desktop Publishing.

Users can consult the index to learn all of the basic terms associated with these categories of specific interest.

As computer technology enters every home, library, school, and office, a new language is developing and growing that will shape and inform everyday life in the dawning "Information Age." The *21st Century Dictionary of Computer Terms* is an indispensable tool for users of all ages and expertise as they learn and apply this vital new language.

A

abend The acronym for ABNORMAL ENDING, referring to the termination of a function or program before it is completed. This generally results from an error condition caused either by the user or by a problem in the hardware or software.

abort The process of ending a function, such as saving a file, before it is complete.

abstract A brief summary of an article, book, or other document.

AC The abbreviation for ALTERNATING CURRENT, used to describe the electricity transmitted through the standard wall outlet.

1

accelerator board An inexpensive way to expand the power of a personal computer, without buying a new system, by adding a faster CPU and, generally, more memory. Accelerator boards come in many variations.

acceptance test One of the stages in the process of designing a software application. During the acceptance test, users are given the opportunity to test the software application as they perform their day-to-day job functions, and then to decide whether or not to accept it, depending on how usable and reliable it is.

access A broadly used term generally meaning to locate data. It can also mean to make use of a computer, a computer program, or a device like a printer.

access arm The mechanical arm in a disk drive that moves the device that reads or writes data to the correct location on the disk.

access code A code that admits a user into a computer system or to a physical location guarded by a computer-based security system.

accessory A device, connected to a computer, that performs related functions, such as a printer or modem.

access time The amount of time required to locate information in the computer's storage and make it available to the user.

accounting software Computer financial programs offering capabilities ranging from personal money management to sophisticated systems for corporations and banks.

accuracy When used in computer science, the overall reliability, or trustworthiness, of the data that results from a computerized operation.

ACM The abbreviation of ASSOCIATION FOR COMPUTER MACHINERY, an organization for technical professionals with an extensive educational and publications program.

acoustical sound enclosure A plastic enclosure placed around an impact printer to help muffle the sound that it generates during the printing process.

acoustic coupler A device that enables a computer to be connected to a network by using a telephone handset. Most acoustic couplers have been replaced by modems, which offer a direct connection.

ACP The abbreviation of ASSOCIATE COMPUTER PRO-
FESSIONAL, a certificate awarded by the Institute
for Certification of Computer Professionals.

ACPA The abbreviation of ASSOCIATION OF COM-
PUTER PROFESSIONALS AND ANALYSTS, a professional organization.

acronym A convention used often in computer science in which multiword terms are shortened by
taking the first letter of each word.

action The execution of a process by the computer.
For example, when the computer responds to a
command from the user to send a document to
the printer, it is performing an action.

action oriented management report A report
generated by a software product that monitors
functions on the computer, such as security, and
provides recommendations for specific actions
to be taken.

action statement A command in a computer program or software product that indicates to the
computer that an action should be performed.

active Currently in use, as is a computer program
that is currently running.

active file The file currently in use. For example, a word processing document currently being written or edited is an active file.

active-matrix LCD A type of liquid crystal display technology, generally used on laptop and notebook computers, in which the screen is refreshed more often than conventional liquid crystal displays. This enhances the user's ability to read what is on the screen.

ACU The abbreviation for AUTOMATIC CALLING UNIT. A device contained in a piece of equipment, such as a facsimile machine, that enables it to be connected to a telephone line.

Ada A programming language used primarily by the U.S. military.

adapter A device, generally a printed circuit board, that makes it possible to connect a specific peripheral unit device, like a graphics monitor, to a computer.

adapter board An adapter in a printed circuit board format.

ADB The abbreviation for APPLE DESKTOP BUS, a built-in interface on Apple computers that allows for the connection of input devices like keyboards.

add-in Components, like increased memory, that can be added to a computer, generally onto an already-installed printed circuit board.

add-on A synonym for ADD-IN.

address A location in a computer's main memory in which data is stored. Addresses can be expressed as numbers, words, or symbols.

address bus A path for moving address data between locations in a computer system.

address space The amount of space available to a specific computer application, represented by a range of addresses.

aggregate function A computation on a group of values rather than a single value. Finding an average among a set of numbers is an example of an aggregate function.

AI The abbreviation of ARTIFICIAL INTELLIGENCE.

alert box Also referred to as a MESSAGE BOX, a space on the monitor screen that displays messages and warnings about the potential effect of an action the user is about to take, such as deleting information.

algorithm Very specific rules or steps prescribed for solving a problem. Computers use algorithms to "think," perform operations, and provide answers.

alias An alternative for FILE or COMPUTER DEVICE, used for the purpose of an application.

aliasing On many computer monitors, the slightly jagged appearance given to perfectly curved lines when the resolution is not high enough.

allocate To provide a user with a computer resource, such as a printer or an amount of main memory, for a specific use.

allocation The process by which resources are assigned.

alphabetic Data, consisting of letters or symbols, listed by letter of alphabet rather than numerical sequence.

alphabetic string A group of alphabetic data unbroken by blank spaces.

alphanumeric Data consisting of letters, numbers (0–9), and, in most cases, symbols.

alphanumeric sort A computer-generated sort of data in either alphabetic or numeric order, or both.

alternate routing The process of moving data along a different path, to get them from one system to another, when the original path becomes unavailable.

Alt key An abbreviation of ALTERNATE KEY, which is generally used within specific applications programs, such as word processing packages, to perform certain functions.

ALU An abbreviation of ARITHMETIC LOGIC UNIT, the portion of the computer used for arithmetic and logical operations.

American National Standards Institute An organization involved in developing voluntary standards for the computer industry. Members represent large technology companies.

American Society for Information Science A professional association for individuals, such as librarians, who work in the field of information science.

American Standard Code for Information Interchange   Generally referred to by its acronym, ASCII (pronounced "as-key"), a standard code that facilitates communication between various computers and computer programs. Most word processing programs allow the user to store text files in ASCII format, so that a document developed under one word processing program can be retrieved and edited within another word processing program.

Amiga   Commodore Computer's line of personal computers with extensive graphics and sound capabilities.

analog   The way in which a variable, such as time, is represented. Analog is continuous, as contrasted to digital, which is finite. For example, an analog watch has a standard clock face with twelve hours and hands that sweep around to show the current time and the intervals in between. A digital watch is finite and shows only the current time.

analog computer   A computer designed to provide continuous measurements of variables such as environmental conditions.

analog monitor   The standard color display monitor used with most personal computers, in which signals are accepted at a specific frequency.

analysis　Using formal (industry- or companywide) techniques to investigate needs and problems as a means of diagnosis and system development.

analyst　An individual who applies formal procedures to the investigation of a problem. Analysts, also referred to as SYSTEMS ANALYSTS, are involved in studying business problems and developing computerized solutions.

AND operator　A term in BOOLEAN LOGIC used in various data retrieval and programming languages. When an AND operator is placed between two statements, it is assumed that the statements on both sides of the AND operator are true.

animation　Using the capabilities of the computer to make an object appear as if it were moving by replicating the same image in slightly altered positions and displaying the resulting series in quick succession.

annotation　Notes or comments, from the computer programmer or software developer, added to computer programs or flowcharts. These notes can later be read by other developers to help them understand the original intent and logic.

annotation symbol Symbols used in annotation, particularly in flowcharts. Examples include boxes and triangles.

ANSI (See AMERICAN NATIONAL STANDARDS INSTITUTE.)

ANSI.SYS A programming term, in DOS and OS/2 systems, that refers to a programming code that adds ANSI compatibility to computer programs.

answer-only modem An inexpensive modem that allows the user to receive but not send data.

antialiasing A procedure in computer graphics for smoothing the jagged lines that often appear in rounded shapes on a low resolution monitor. (See also ALIASING.)

antistatic mat A floor mat placed in front of a computer or other sensitive device to prevent static electricity that can result in loss of data.

API (See APPLICATION PROGRAM INTERFACE.)

app An abbreviation of APPLICATION.

append To add to the end of something. For example, a user might add, or append, a new item at the end of a list.

Apple Desktop Bus An interface built into Apple Macintosh computers that allows devices, such as keyboards, to be attached.

Apple Macintosh Computer A popular brand of personal computer produced by Apple Computer, Inc., with features that include extensive graphics, ease of use, and input through a mouse and keyboard.

AppleTalk Networking capabilities provided in Apple Macintosh computers that allow them to communicate with each other over a local area network.

application A program that enables users to employ the computer in the performance of daily job functions. Applications have been designed for diverse industries and functions, with examples that range from engineering design, airline reservations, and banking, to word processing and desktop publishing. Application programs are developed for both mainframe and personal computers.

application program Computer programs designed to enable users to perform specific job functions. Word processing, accounting, and engineering programs are examples of application programs.

application program interface Code in one computer program that allows it to work with others. As a result of application program interfaces, users can use one set of procedures to operate a variety of different programs.

application programmer A computer programmer with a focus on developing specific user applications, such as programs used in finance or word processing.

application programming The process of developing application programs.

application software (See APPLICATION PROGRAM.)

applied mathematics The use of mathematical concepts and principles in day-to-day activities. Computer science is an example of applied mathematics.

ARC A format for compressing files (see also DATA COMPRESSION).

arcade game   A computer game modeled after the pinball machine concept, in which joysticks manipulate figures and icons around the computer screen. Arcade games use animation and sound.

architecture   A broadly used term that can refer to the design of software, hardware, or a complete computer system. Generally, computer manufacturers adopt a certain design, or architecture, and then build other computers and software to work within the constraints of this design.

archival backup   A long-term method of backing up data by storing old files on media such as tape, rather than using valuable space on a disk. Files that have been backed up archivally will take longer to restore without careful records of which tape or other medium the files are stored on.

archive   The process of backing up or storing files with the archival backup method. ARCHIVE also refers to the area in which these data are stored.

ARCnet   A type of local area network, known for its simplicity and ease of use.

argument   A programming routine; a variable to which different types of values can be assigned.

arithmetic expression A combination of symbols, used in programming, that represents a numeric value.

arithmetic operator A symbol, read by the computer, that represents an arithmetic action, such as addition, subtraction, multiplication or division.

ARPANET A communications network established by the Advanced Research Projects Agency of the Department of Defense for companies and universities involved in defense projects.

array A series of items related in shape or size.

arrow keys Keys on a computer keyboard that move the cursor in different directions.

Artificial Intelligence A branch of computer science focused on how computers can be programmed to think and reason like human beings.

artificial language A computer programming language, such as COBOL, with its own set of rules and syntax.

ASAP The abbreviation of AS SOON AS POSSIBLE.

ascending order Arranging items in order from first to last or smallest to largest.

ASCII (See AMERICAN STANDARD CODE FOR INFORMATION INTERCHANGE.)

assemble In the context of computer programming, to translate symbols into code the computer can read.

assembler A computer program that assembles, or translates, symbolic code into instructions usable by the computer.

assign To give a value to a variable, such as giving a specific number to a variable in a computer program.

Association for Computer Machinery (See ACM.)

Association for Computers and the Humanities An association dedicated to understanding the relationship between computer science and fields such as language, literary studies, art, music, and dance.

async An abbreviation of ASYNCHRONOUS.

asynchronous Communications in which data are not transmitted at regular intervals, such as a telephone conversation.

asynchronous device A device that transmits communications signals at irregular intervals.

asynchronous transmission The process of transmitting data at irregular intervals with symbols that indicate the beginning and end of a signal.

AT An abbreviation of ADVANCED TECHNOLOGY, used to refer to an IBM personal computer with an Intel 80286 microprocessor and a 1.2 MB floppy disk drive.

AT bus The expansion bus on IBM AT and compatible personal computers, making it possible to upgrade AT systems with additional memory and add other devices. (See also BUS.)

ATM An abbreviation of AUTOMATIC TELLER MACHINE, a sophisticated terminal that allows users to complete banking operations by interacting directly with the bank's computer system, without human intervention.

attach To connect a peripheral device, such as a printer, to a computer system, or to connect a computer to a network.

attribute A characteristic of data, a data field in a database, or a system. For example, an alphanumeric field in a database has an alphanumeric attribute.

audio Audible; able to be heard. Multimedia programs for personal computers, including arcade games, often include audio capabilities.

audio device A device on a computer that transmits sound.

audio output Sound output from a computer in the form of music, synthesized human voice, or other sounds.

audio response device A computer device that transmits a synthesized human voice.

audio-visual Materials that go beyond standard printed literature to include color illustrations or sound.

audit trail A method of tracing the actions in a software system over a period of time. For example, an audit trail in a mainframe computer environment might indicate when users are logging on to the system, when they are using the system the most, and what kinds of tasks they are performing. An audit trail might also indicate when an unauthorized user is attempting to break into the system. Audit trails are produced by software products that collect and organize input and output data.

author   An individual who designs computer-based training courses.

authoring system   A software product that simplifies the process of designing and writing computer-based training courses, often including multimedia capabilities.

authoring tool   Another term for AUTHORING SYSTEM.

authorization   Permission to access restricted data, application programs, or specific features within an application program. For example, only personnel executives have authorization to access confidential employee records.

authorized program   A computer program capable of making major changes in the operation of a computer system.

author language   The language used in designing, or authoring, computer-based training programs. Author languages based on menus and simplified commands are generally relatively easy to use.

auto-answer   A feature on a modem that automatically answers incoming calls, regardless of whether the user is present.

auto-dial   A feature on a modem that allows the user to automatically dial telephone numbers by, for example, clicking an icon, using a code, or choosing an item from a menu. Special communications software is necessary for this feature.

auto-dialing modem   A modem that offers the auto-dialing feature.

autoexec.bat   The abbreviation of AUTOMATIC EXECUTE BATCH FILE, a DOS file. Autoexec.bat executes automatically during the process of starting up (see BOOTING) a DOS-based personal computer. Often-used programs, such as a word processing program, may be loaded automatically at system startup by including commands in the autoexec.bat file.

automated office An office environment in which technology is extensively in use. This might include various computers and applications such as word processing and electronic mail, as well as telecommunications capabilities.

automatic In the computer context, procedures that are executed without human intervention.

automatic calling unit (See ACU.)

automatic carriage A printer feature that automatically feeds paper to be printed and ejects finished documents.

automatic error correction A feature available with certain technology products that detects and corrects standard errors without intervention by the user.

automatic programming Using special software programs that simplify the process of developing computer program code. For example, products are available that allow users to develop a computer program through the use of icons without having to learn a specific programming language.

automatic scrolling Going forward or backward through a document on the computer screen,

21

line by line, generally by pressing the up or down arrow key. Also referred to as CONTINOUS SCROLLING.

automatic shutdown   The ability of a computer system to end operations on its own, such as at a predetermined time at the end of the work day.

automatic teller machine   (See ATM.)

automation   Controlling procedures and operations by computer, without human intervention; for example, a computer system that controls room temperature and lighting.

automaton   A robotlike machine that performs some human tasks.

automonitor   A computer software program that watches over the operations of the computer system, checking for hardware and network problems, and recording data on the system's overall performance.

autopolling   A feature on facsimile machines, as well as computer networks, in which other connected devices are periodically scanned to determine whether they can send or receive information.

auto-redial A modem feature that repeatedly re-dials a busy telephone number until a connection is made.

auto-repeat A computer keyboard feature that repeats a character when the key is held down. For example, if the space key is held down on most computer keyboards, the cursor will continue to move across the screen.

auto-restart Generally a mainframe term referring to the automatic restarting, or REBOOTING, of a computer after it has discontinued operation. A computer that automatically begins its standard restart procedure after shutting down in a power outage is an example of auto-restart.

autosave A feature often incorporated in word processing programs in which data are periodically saved without user intervention. For example, a user might designate the feature to automatically save the current document at five-minute intervals.

autosizing In computer graphics, the automatic adjustment of images to fit the size of the computer screen.

autotracing In computer graphics the automatic conversion of images from raster to vector format (see RASTER and VECTOR) through the graphics software program.

auxiliary memory Storage supplemental to the main storage of the computer. Examples of auxiliary storage include CD-ROM and floppy disks.

auxiliary operation Operations that are performed outside the main computer processing unit. Printing a document that has been sent from the computer to a printer is an example of an auxiliary operation.

auxiliary storage Another term for AUXILIARY MEMORY.

availability The time that a computer system, or specific applications, are actually in working order and accessible to the user community.

AWC The abbreviation of ASSOCIATION OF WOMEN IN COMPUTING, a professional organization of women computer science professionals.

B

b An abbreviation used either for BIT, if referring to data storage, or BAUD, if referring to data communications.

B The abbreviation for BYTE.

B: The second floppy-disk drive in a personal computer using the DOS operating system.

babble A condition in which communications channels on a system interface with each other, so that some of the signals carried on one channel are also carried on another. (See also CROSSTALK.)

backbone The wiring system that connects the various nodes on a computer network.

back end In programming slang, the last section of a computer program. Also, computer hardware the average user is not involved with, such as the hardware that performs database searches.

background On a monitor, the blank areas of the screen where characters are not displayed. Characters and graphics are in the FOREGROUND of the screen.

background noise Interference on a communications line from other lines.

background processing Lower-priority processing that occurs when the computer is not currently otherwise engaged, such as performing a minor database operation (like producing a mailing list) when the user is not actively retrieving data.

background program A computer program capable of running as a low priority while other, higher-priority programs are in operation. An example is a "screen saver" program, which produces graphic images on the screen when the system is idle for a period of time.

background task A low-priority operation processed when the system is not currently otherwise engaged (see also BACKGROUND PROCESSING).

backing up The process of saving copies of documents and other data so that, in the event of a system failure or other emergency, this information is not lost. Data from a hard disk are often backed up onto floppy disks.

backlighting A technology often used with notebook computers to help make the flat-panel display more readable. A backlit screen is lighter than the characters that are displayed on it.

back panel The back a computer cabinet, containing outlets or sockets used to connect devices like a printer or monitor.

back-slash The back-slash symbol, \ , on the standard keyboard is used, in DOS, in the process of designating directories and root directories. DOS-based programs, such as word processing packages, also use the back-slash in creating document directories.

backspace key The key used in backspacing.

backspacing Using the backspace key on keyboard to move the cursor to the left, one space at a time. Generally, this process results in erasing whatever symbols have been typed.

back up The process of making additional copies of data to protect them from unexpected disaster.

back up and recover Generally, the process of assuring that valuable data and application programs stored in the computer have also been recorded on another medium so that, in the event of a disaster, they are not lost.

back up and restore Another term for BACK UP AND RECOVERY.

backup copy A copy of a document or other data stored for use in the event of unexpected destruction of the original.

backward compatible Used by software developers to indicate that a new version of their product can use data created by an older version. For example, a new version of a spreadsheet product can generally use data that was created and stored by an older version of the product.

badge reader A device that can read security badges with special codes, as well as the magnetic tape on the backs of credit cards.

bad sector Disks, both hard and floppy, are divided into sectors that facilitate the storage of data. A bad sector is somehow damaged and cannot be used for data storage.

band A defined spectrum of frequencies across which data can be sent. BAND is used within terms such as BANDWIDTH.

band printer A printer that operates with characters embossed on a band, usually made of steel. Band printers do not generally produce the high-quality output associated with laser printers, but they are fast and are thus often used for high-speed printing such as customer invoices.

bandwidth In data communications, the capacity of a band for transmitting data. The higher the bandwidth, the more data that can be transferred.

bar chart A graphic representation of data using rectangular bars to show relationships among different values. Bar charts are used in business and research applications.

bar code A code used in representing data, consisting of parallel bars that can be read by a scanner and used as input into a computer. Bar codes are often used in grocery stores for product and price information, and also in hospitals and libraries.

base address Generally refers to the beginning of a computer program. The base address is the starting point for other addresses (see also ADDRESS).

baseband transmission A low-frequency transmission in which one signal at a time is carried, as over a local area network, where all computers on the network are only a short distance from each other.

BASIC The acronym for BEGINNER'S ALL-PURPOSE SYMBOLIC INSTRUCTION CODE, a relatively simple programming language, with a limited number of commands, most often used in programming personal computers.

Basic Input/Output System Basic, "built-in" code on a personal computer that works independently of the operating system, such as DOS, and controls the keyboard and other parts of the computer. Basic Input/Output System is also referred to as BIOS.

batch A group of records treated as a single unit for computer processing. For example, a group of payroll or personnel records might be processed as a batch.

batch file A file that contains a group of commands used in batch processing.

batch file transmission Sending a group of files in one transmission, using a single command.

batch job A task or operation processed without any involvement from the user, often after the normal workday when there are fewer demands on computer resources. An example of a batch job is printing a lengthy mailing list.

batch processing Completing tasks that entail the use of the same computer program with groups of records, or batches. Batch processing usually occurs during off-hours. For example, banks update the deposits and withdrawals of their depositors at night, as a batch job, when the computers are not needed for other tasks. Also, payroll checks are usually processed as a batch job, with a check printed for each employee, during off-hours, for efficient use of the computer.

batch program A computer program designed to process tasks without requiring user involvement.

batch system A system (hardware, operating system software, and application software) designed to process groups of tasks without requiring user involvement.

BAT file A DOS term for BATCH FILES.

battery backup Battery power available to operate a computer in the event of a power failure.

baud One bit per second, used in measuring electrical oscillations used for transmitting data.

baud rate The speed at which data is transmitted, generally over a telephone line with the use of a modem. For the average personal computer user, modem baud rates range from 300 to 9600.

bay A place where hardware, such as a floppy drive, can be installed. For example, personal computers often include space where an additional disk drive may be installed.

BBS The abbreviation of BULLETIN BOARD SYSTEM, available to online computer users who wish to post messages and receive responses from other online users.

Bell 103 The standard protocol for transmitting data over telephone lines at rates of 300 baud or less.

bells-and-whistles A slang term for the "fancy" features on a computer or in a software program. Bells-and-whistles might, for example, include sound and color graphics.

Bell 212A The standard protocol for transmitting data over the telephone at a rate of 1200 baud or less.

benchmark A basis for measuring the performance of a computer or software system. A potential user may, for example, want estimates of the speed and capacity of the system when used under certain conditions by a set number of users. To this end, basic figures on the product's performance under specific conditions might be provided by the vendor, or the potential customer might install the product to obtain more specific benchmark figures.

benchmark problem A set of conditions for testing a computer hardware or software product to obtain benchmark figures.

benchmark tests The actual tests for creating benchmarks of computer hardware or software programs, such as speed capabilities or the number of users that can be accommodated.

Bernoulli Box A removable hard disk, manufactured by Iomega Corporation.

Bernoulli Disk Drive Another term for BERNOULLI BOX.

beta test Testing a computer hardware or software product after the development process is completed but before the product is actually put into use by average users on a daily basis. During the beta test, any additional "bugs" will ideally be discovered and fixed.

BFT Abbreviation of BINARY FILE TRANSFER, a standard used in transmitting data from one facsimile modem to another.

bi-directional The ability to send data in both directions.

bi-directional printer Printing technology that enables a printer to print a line across a page in one direction, and then print the next line in the other direction without an additional carriage return. This results in greatly increased printing speeds.

Big Blue   A slang term in the finance and technology industries to refer to IBM, derived from IBM's use of blue as a corporate color.

binary   A number system in which only two digits are possible. Computers use a binary number system based on 0 and 1, and are programmed with a machine code using combinations of these two digits.

binary code   A code using only two digits, 0 and 1.

binary file   A file stored in machine code and thus readable by the computer but not by humans.

binary file transfer   (See BFT.)

Binary Format   Use of a binary code to represent data, particularly numeric data.

BIOS   (See BASIC INPUT/OUTPUT SYSTEM.)

bipolar   An integrated circuit consisting of two layers of silicon.

BIS   The abbreviation of BUSINESS INFORMATION SYSTEM, broadly used to refer to computers, peripherals, software, and data designed to work together to meet the needs of a business organization.

B-ISDN BROADBAND ISDN, a standard that employs fiber optic lines different types of data, including video and voice.

bit An acronym for BINARY DIGIT. A bit is composed of either a 0 or a 1, and is the smallest unit of information used on a computer. A BYTE, a word recognized by a computer, is composed of a group of bits.

bit map In graphics technology, when a portion of the computer's memory is used for storing graphics. This portion consists of dots represented by bits and organized in columns and rows. The greater the number of bits used in representing a dot, the greater the resolution that results.

bit-mapped font A font in which each character is composed of pixels that represent a specific pattern of bits stored in memory.

bit-mapped graphics Graphics that are stored as a bit map. Bit-mapped graphics are also referred to as RASTER GRAPHICS.

BITNET A large, wide-area network for accessing and sending information among geographically dispersed users; often used by universities.

bits per inch The number of bits that can be stored in an inch of space on a storage medium, such as a tape.

bits per second The number of bits that can be transferred, over a telecommunications line, in a second of time. Modems, which use telephone lines, transfer data at ranges of 300 to 9,600 bits per second. (See also BAUD RATE.)

black box An electronic device that either scrambles or decodes electronic signals.

blank An empty space, such as that occurring between two words. Also refers to the process of erasing—blanking—a computer screen.

block A word processing term for a group of words that have been marked so that a special operation can be performed on them. For example, a paragraph can be blocked using function key. This paragraph, or block, can then be moved to another part of the document, or to another document entirely.

block diagram An illustration showing the elements of a computer system, with each element represented by a separate block and lines connecting the blocks to show the relationships between elements.

blocking The process of designating a group of words as a block. (See BLOCK.)

block move The process of moving a block of words from one position in a document to another, generally using function keys.

board A printed circuit board that consists of chips and wires and is inserted in the computer to enable the performance of tasks such as communications and graphics use. (See also CARD.)

boilerplate Sentences and paragraphs used repeatedly in different documents, such as press releases and proposals.

boldface Words printed much darker than the other words in a document.

bomb A failed computer program. Also, the process of failing.

Boolean expression An expression in which a comparison is made and the result is either true or false. For example, $7 + 8 = 15$ is a Boolean expression. The term derives from its inventor, George Boole.

Boolean logic   The underlying logic behind Boolean expressions. Boolean logic is used extensively in programming as well as in fourth-generation languages.

Boolean operator   The operators generally used in Boolean logic, including AND, OR, and NOT.

boot   Starting up a computer; specifically, the process of loading the operating system into memory.

boot block   A small section of a hard or floppy disk containing the programs needed by the computer during the start-up process.

bootstrap   The process undertaken by a computer "booting itself up," including loading the operating system into main memory and making the graphical user interface available.

bootstrap loader   The program used by the computer to begin the start-up process, including checking the memory and connections to devices.

bottleneck   The result of overloading the processing capacity of a computer with so many commands that its input/output devices are clogged with data, and work is either greatly slowed or comes to a complete standstill.

box   A slang term for computer.

bpi   The abbreviation of BITS PER INCH, the number of bytes that can be stored on an inch of magnetic tape.

bps   The abbreviation of BITS PER SECOND. (See also BPI.)

break   To stop an operation while it is being performed, or a data transmission while it is in process.

break key   A key that can be pressed to interrupt the operation currently being performed by the computer. Generally, the break key is available on keyboards attached to mainframe computers.

brightness   The overall level of luminosity of the characters, or of colors, on a screen. Brightness is not readily measurable and is a function of perceptual differences among individuals.

broadband   In data communications, a network that can carry a variety of signals at the same time; often used for local area networks. Broadband networks are generally based on coaxial or fiber-optic cables.

Broadband ISDN   (See B-ISDN.)

broadband transmission   Data communications that require the use of broadband transmission facilities.

broadcast   Transmitting data to a number of receivers at the same time. An example of broadcasting is sending the same message to a number of facsimile machines by a special broadcast facility that executes this multiple transmission with a single user action.

browse   The process of looking at information stored in a database, for example, or the contents of documents in a word processing program.

brush   An onscreen instrument in software programs that allows the user to fill in images with color—to "paint" the images. Users can choose the type of brush desired and manipulate it with a mouse.

buffer   A temporary storage area in transmitting data from a computer to another device. Buffers are often used with printers so that a large amount of data can be sent to a printer and stored in a buffer area while waiting to be printed. This avoids tying up the system and allows the user to go on to other work.

bug   An error in a computer program.

built-in font   A font provided standard with a specific printer. Built-in fonts are essentially part of the printer's hardware, while other fonts may be added through software.

bulk eraser   A device to quickly and completely delete the data currently stored on a floppy disk, rather than having to delete one file at a time.

bulk storage   A device, such as a disk or tape, that is capable of storing large amounts of data.

bullet   A mark, such as a line or a large dot, to separate the items in a list.

Bulletin Board System (BBS)   Networks that users may dial into, through the use of a computer and modem, to communicate with other users. BBSs are usually organized around specific interests, with members leaving messages for each other and sharing information and software.

bundle   A marketing technique in which a product, such as a computer or software package, is sold with other products included. For example, personal computers are often bundled with an operating system and a graphical user interface.

burn-in   Allowing the same image to remain on the screen for long periods of time, so that a shadow of the image is etched permanently into the screen's phosphorus coating. Burn-in can be distracting to the user and will interfere with the clarity of screen images.

bus   Wires to provide a route for transmitting data. The bus connects the computer to other devices, such as the printer.

business graphics   Graphics useful in business functions, such as sales presentations. Business graphics include pie charts, histograms, and other diagrams that illustrate concepts such as demographic trends and profitability over a period of time.

business information system   (See BIS.)

business software   Application software designed to perform business-related functions, including personnel and finance.

bus mouse   A mouse connected to the computer through an expansion board rather than a serial port. This frees the serial port to be used for another device, such as a modem.

bus network   The use of a common standard bus to enable communication between different computers.

bus topology   A technique for connecting the computers in a local area network in which all computers on the network are connected through one central cable.

buzzword   Terms of slang and abbreviations used by individuals in a specific profession. The computer industry is rife with buzzwords such as GUI (pronounced "gooey," for GRAPHICAL USER INTERFACE).

byte   A group of eight bits. The computer recognizes a byte as a unit, or a single character.

bytes per inch   (See BPI.)

C

C A relatively new computer programming language, developed during the mid-1970s. C has obtained a high level of popularity because it is easy to use, compared to other languages, and is useful in a wide range of applications. C can be used on the personal computer and with the UNIX operating system.

C++ A version of the C programming language with object-oriented capabilities added.

cable A wire used to connect the components of a computer system, for example, to connect a printer to a personal computer.

cable connector   The plugs used to connect cables, located on devices like printers and personal computers.

cabling diagram   An illustration that provides an overview of physical connections between computers and other devices in an organization's system.

cache memory   A mechanism that provides high-speed memory. Cache memory provides faster access to data by serving as a temporary storage area between the main storage area, where access to data can be slower, and the central processing unit. Cache is pronounced "cash."

CAD   The abbreviation of COMPUTER-AIDED DESIGN, a computer base system used in industrial design—anything from equipment to buildings. CAD involves the use of either sophisticated personal computers or workstations, with specialized software that generally incorporates state-of-the-art graphics capabilities. CAD essentially automates the function performed by the drafter.

CADAM   The abbreviation of COMPUTER-GRAPHICS AUGMENTED DESIGN AND MANUFACTURING, in which the computer is used for designing and producing various manufactured products. CADAM is closely aligned with CAD.

CAD/CAM The abbreviation of COMPUTER-AIDED DE-SIGN/COMPUTER-AIDED MANUFACTURING, and similar to CADAM, based on the premise that the computer can be used throughout the production process in a factory, from initial design through the final stages of manufacturing.

CADD The abbreviation for COMPUTER-AIDED DE-SIGN AND DRAFTING, referring to CAD systems with special features that automate drafting functions such as the process of adding notes to a drawing.

CAE The abbreviation of COMPUTER-AIDED ENGI-NEERING, in which drawings from CAD or CAD/CAM systems form the basis of engineering functions, including testing how a product will work under simulated conditions. Many CAD and CAD/CAM products offer CAE functionality.

CAI The abbreviation of COMPUTER-ASSISTED IN-STRUCTION, in which computers form the basis of the instructional process. For example, students may be presented with computer-based lessons to allow them to work individually and at their own pace. Once a lesson is completed, the student may be presented with questions to test what has been learned and either guided through additional materials or directed to proceed to the next lesson.

CAL The abbreviation of COMPUTER-AUGMENTED LEARNING, in which computers supplement the materials presented in the classroom. For example, students may be directed to complete additional exercises or simulations on the computer to provide additional practice and instruction.

calculator A machine, generally hand-held, that performs mathematical calculations, or a software program that provides the same capabilities.

calendar A software program that can be used to keep track of a personal schedule; it may also include features like "pop-up messages" to serve as a reminder of important events and deadlines.

call In programming, when control of the program is passed to a subroutine. For example, in a checkbook-balancing program, a negative value in the total column might result in a call to a subroutine that prints the message: "Your checking account is overdrawn."

callback modem A more secure form of modem that does not allow the caller instant access. Instead, the caller must enter a code when making contact with the callback modem. This code is checked in a database, containing codes and associated callers and their modem telephone numbers, before the callback modem returns the call to provide access.

CAM The abbreviation of COMPUTER-AIDED MANU-FACTURING. (See CAD/CAM.)

camera-ready Pages of a book, newsletter, or other document, produced through a desktop publishing system, that are at the stage where they can be sent directly to an offset printing service to be photocopied and made into plates for printing.

cancel A command that halts an action currently being performed, restores a previous deletion, or erases the line on which the cursor is currently positioned. The actual result of using the cancel command depends on the software program being used.

canned software Software purchased from a vendor, as opposed to custom software, which is developed from scratch. A spreadsheet or word processing package purchased from a computer software store or out of a catalog is an example of canned software. Another term for canned software is OFF THE SHELF.

capacity The amount of data that can be stored on a storage medium, such as a floppy disk. Also the number of operations a computer can process at one time.

Caps Lock Key A key on most keyboards that, when pressed, causes all letters to appear in upper case. When pressed again, subsequent letters appear in lower case.

capture Another term for saving data in a computer.

carbon ribbon A high-quality ribbon made of Mylar and used in dot-matrix printers.

card A printed circuit board that provides functions such as communications or graphics, placed in slot inside a personal computer (see also BOARD).

card reader A device that reads data encoded on a magnetic strip; used, for example, on credit cards and bank machine cards.

carriage A mechanism on a dot-matrix or band printer that feeds paper into the printer.

carriage return Abbreviated CR, a command in a word processing program that causes the cursor to jump to the beginning of the next line. In a document, a carriage return is used to begin a new paragraph. In many word processing programs, the Enter key is used to create a carriage return.

carrier   A company that provides telecommunications or data communications services.

cartridge   A unit that can be inserted into a slot in a computer to perform special functions. A cartridge may contain a storage medium, such as a tape, or it may contain software stored in ROM. Many of the software programs available for early personal computers were stored on cartridges. Laser printers also make use of cartridges, on which additional fonts are stored.

cartridge font   A cartridge containing fonts, for use with a printer.

cascading windows   Windows in a graphical user interface, displayed in an overlapping arrangement.

case   Used to indicate whether characters, in a program or document, for example, are capitalized (uppercase) or not (lowercase).

CASE   The acronym for COMPUTER-AIDED SOFTWARE ENGINEERING, broadly used to mean software that assists in the process of developing application programs. CASE software can assist in any or all phases of the development process, from initial design to coding to final testing.

case sensitivity   Refers to the need to use either upper- or lowercase commands in a computer program. Case-sensitive programs respond differently to commands entered in upper-case and in lower-case.

cassette tape   Magnetic tape for data storage, enclosed in a plastic cassette.

CAT   The acronym for COMPUTER-AIDED TESTING, used to describe a category of software that assists programmers in testing application programs for design flaws.

catalog   A list of files stored on a disk. For example, the list of documents created and stored through a word processing program is a catalog. Generally, catalogs include names of files as well as brief information about the size of each file.

cathode-ray tube   The technical term for a monitor or computer terminal. Refers to the technology on which monitors, as well as televisions, is based.

CBEMA The abbreviation of COMPUTER AND BUSINESS EQUIPMENT MANUFACTURERS ASSOCIATION, a professional association representing the hardware industry and involved in setting standards.

CBL The abbreviation of COMPUTER-BASED LEARNING, broadly used to refer to the use of the computer in an education setting.

CBT The abbreviation of COMPUTER-BASED TRAINING, broadly used to refer to the use of the computer in an education setting.

CCITT The abbreviation of COMITE CONSULTATIF INTERNATIONAL TELEPHONIQUE ET TELEGRAPHIQUE. This is an international organization that develops standards for communicating wth facsimile machines, modems, telephones, local area networks, electronic mail, and other media.

CD-ROM The abbreviation of COMPACT DISC–READ-ONLY MEMORY, with READ ONLY referring to the ability to "read" data from the disk, but not "write" on the disk. With CD-ROMS, information is contained directly on the disc and accessed with a microcomputer and CD-ROM player. One compact disc has enough storage space for the equivalent of 275,000 double-spaced typed pages. CD-ROM has found its niche in multimedia, where CD-ROM is used for storing, not only text, but full-color images and sound.

CD-ROM drive A device, attached to a computer, for retrieving data from CD-ROM disks.

CE The abbreviation of CUSTOMER ENGINEER, a representative from a software or hardware company who assists customers in using the company's products. CEs often work at customer sites for specific periods of time, or even long-term.

cell A single box on a spreadsheet grid, which is a matrix of rows and columns. Each value on a spreadsheet is entered in a cell.

centralized data processing The traditional approach to data processing, in which one or more large computers, under the control of the information services department, serve the data processing needs of users and departments throughout the organization. This can be contrasted with the decentralized approach, in which the work is performed by smaller computers spread out among various locations.

centralized design The structure underlying the centralized data processing approach to computing.

centralized processing (See CENTRALIZED DATA PROCESSING.)

central processing unit The basic element of a computer, where commands are executed; synonymous with the computer itself. Also referred to as the CPU.

central site The physical location of the main computer, also referred to as the GLASS HOUSE.

certification A stamp of quality awarded to a hardware or software product by an organization that has been empowered to make this decision. Certification is also awarded to individuals who have met specific criteria established by a professional organization.

CGA The abbreviation of COLOR GRAPHICS ADAPTER, the earliest graphics systems for IBM personal computers. CGA has relatively poor resolution and a minimal number of colors (16), and has since been replaced by other, more sophisticated systems such as VGA and EGA.

chain A database concept in which related database records are linked together, one after the other.

chaining The process of linking database records together.

change file A file in a database management software program that records all retrievals of, and changes to, the contents of the database. The change file records all database activity, particularly in the event of an error or a security violation.

channel A path, or link, between two computers.

channel adapter A device that makes communication possible between different hardware devices.

channel capacity The speed at which data can be transmitted across a communications line; generally measured in BITS PER SECOND or BAUD RATE.

character Letters, numbers, symbols, and punctuation marks that can be stored in a computer. A character requires one byte of storage.

character-based Programs that can handle only standard characters as opposed to graphics. Word processing and spreadsheet programs are examples of character-based programs.

character checking Checking the content of a field in a database for spelling, or to make sure the correct combination of, for example, alphabetic or numerical characters has been entered.

character density The number of characters contained within a specific amount of space—usually an inch—on a screen or printed page. Character density (See also CHARACTER PITCH) is often measured as charcters per inch.

characteristic An attribute of data, a program, or a device.

character mode A graphics format in which the screen can only display characters. Character-based programs will run in character mode, with the screen essentially divided into blocks, and each block containing one character.

character pitch The number of characters in an inch of text. Sophisticated printers can print characters in a variety of sizes; the larger the pitch, the fewer characters per inch.

character recognition The ability of a computer to recognize and understand the characters—numbers, letters, and symbols—that humans use. Without this ability, communication with a computer would require the use of binary code.

character set The list of characters that a specific computer understands. Computers used in the U.S. understand a different character set than those used, for example, in Japan.

characters per inch The number of characters that can be printed in a line of text. (See also CHARACTER PITCH.)

characters per second The average number of characters that can be printed in a second. This measurement is generally used with dot-matrix printers, whereas more sophisticated printing technology, such as laser printing, is measured in pages per minute.

character string A group, or series, of characters recognized by the computer as such.

character style The general attributes of a character, such as whether it appears in boldface or in italics. Style is one aspect of a font.

chart A graph, showing the relationships between various factors.

chassis The metal frame that serves as the basis for a computer. Drives, circuitry, and expansion slots are all mounted on the chassis, which is then covered by the cabinet of the computer.

check sum Assigning numeric values to computer operations, such as sending data. These numeric values can then be summarized and compared with each other, by a computer program, as a means of error detection. If the sum of the transmitted messages and received messages is not equal, an error has occurred.

chiclet keyboard Small, flat keys used on early personal keyboards; derived from the chewing gum of the same name.

chief programmer The individual in charge of other programmers. The chief programmer ensures that all code is developed using the same standards, and that it is developed on time.

child Often, a file contained within a specific directory/subdirectory (parent) on a hard disk.

chip An electronic component, often made of silicon, on which large amounts of information can be stored. Chips can also perform tasks such as arithmetic. The foundation of a computer is chips attached to a printed circuit board.

CICS The abbreviation of CUSTOMER INFORMATION CONTROL SYSTEM—a communications interface, developed by IBM, that allows for the processing of large numbers of transactions. CICS is used with mainframe computers, for handling applications such as bank transactions and airline reservations.

CIM The abbreviation for COMPUTER-INTEGRATED MANUFACTURING, similar to COMPUTER-AIDED MANUFACTURING, in which the computer is used in virtually all aspects of managing a manufacturing facility.

circuit The path along which electrical current flows.

circuit board A thin board for mounting microchips and other electronic components, installed inside a computer functionality. The basic CPU of the computer is contained on a circuit board, called the motherboard, as are circuit boards that provide additional functions, such as graphics capabilities. The boards are often printed with a pattern to indicate how the circuit should flow.

circuit card Another term for CIRCUIT BOARD.

CISC The abbreviation of COMPLEX INSTRUCTION SET COMPUTING, a computer processing unit that handles large numbers of complex instructions and is thus powerful in terms of its capabilities. However, CISC processors are slower than those based on reduced instruction-set computing, or RISC, technology (see definition).

class A group of elements, including data, that are similar to each other.

clear To erase or otherwise cause the screen of a monitor to become blank. Some keyboards include a function that performs this function automatically.

clearing The process of causing a screen to become blank.

click Pressing and quickly releasing the button on a mouse, either once or twice, to perform various functions.

click art Art provided through a graphics product that can be brought into a document in all cut-and-paste operations to add sophisticated graphics without having to actually draw the art from scratch.

clicking Pressing the button on a mouse.

client/server architecture An approach to designing a computer network in which powerful computers act as servers, providing databases, applications, and system management functionality to clients, generally personal computers or workstations. In many organizations using this approach, a server is located in each department, or in each building. Client/server architecture may or may not involve the use of a mainframe computer.

clip art Another term for CLICK ART.

clipboard An area of memory for storing text that is in the process of being moved from one location to another. For example, if a paragraph is being moved to another place in a document, it is held in the clipboard while the user locates the spot where it is being moved.

clipping The process of removing edges, or portions, from a graphics image so that it fits into a smaller space.

clobber To destroy data as a result of an error.

clock An internal device in a computer that maintains the current time and is available to the software products in use. Word processing programs that record the time at which a document was last modified are relying on the computer's clock for this information.

clock speed The speed at which the computer is able to respond to commands given by the user.

clone A look-alike, as when a computer is designed to be a replica of another, using the same software and operating in much the same way. The less-expensive "IBM clone" line of computers is an example.

close In database terms, completing work in a data file and issuing what ever command is necessary (usually close) to indicate that the work is complete and place the new version of the data in storage.

closed architecture Computer hardware or software products that are designed so that other manufacturers cannot design products that work with it. Products based on a closed architecture are referred to as proprietary. Hardware manufacturers have often been criticized for designing products that will not work with other manufacturers' products, thus limiting their customers from readily purchasing other products on the market.

closed file A data file available for authorized users only. Personnel files are closed files.

closed shop An organization's computer system with access limited to individuals with a technical background only, such as programmers, rather than general users.

closed system Another term for CLOSED ARCHITECTURE.

cloth ribbon A ribbon in dot matrix printers, based on an older technology, that does not provide as clear an image as a carbon ribbon.

CMI The abbreviation of COMPUTER-MANAGED INSTRUCTION, a concept similar to computer-based training, in which the computer is an integral part of the educational process.

CMOS The abbreviation of COMPLEMENTARY METAL OXIDE CONDUCTOR, a semiconducter often used in portable computers because of its ability to function with low power levels.

CMS The abbreviation of CONVERSATIONAL MONITOR SYSTEM, an interface to IBM's VM operating system for users and programmers.

coaxial cable A cable often used in local area networks that allows data to be transmitted at high speed.

COBOL An acronym for COMMON BUSINESS ORIENTED LANGUAGE, a programming language used primarily for business applications. COBOL was developed during the 1950s and 1960s and, while still in wide use, has been replaced by more efficient languages such as C.

code Symbols and instructions used to represent words, numbers, and commands, such as those used in a computer program.

code conversion The process of translating program code designed to work on one hardware platform into a form that works on another. An example is rewriting an application program designed to work on an IBM personal computer so that it works on an Apple Macintosh.

coder An individual who performs the step-by-step process of writing a computer program, often under the direction of a chief programmer.

code segment A section of a computer program consisting of a set of instructions designed to perform a specific function. Large programs are often coded in segments, with individual programmers assigned to these specific segments.

coding The process of programming a computer.

coding form   A sheet of paper formatted to guide programmers in handwriting lines of program code, to be typed into the system at a later time. The use of coding forms has declined in favor of working directly on the computer screen.

cold boot   Turning on a computer that is not currently in operation.

cold start   The process of starting a computer that has suffered a failure so serious that it has completely shut down. In a cold start, the computer has to be turned on, and the operating system loaded, just as if it had been sitting idle.

collate   To organize data from various sources into one set of data, as in organizing various lists.

color graphics   The ability of a computer to produce graphic images in color.

color graphics adapter   (See CGA.)

color monitor   A monitor, connected to a personal computer, that displays information in full color. The screens of color monitors are coated with red, green, and blue phosphorus.

color printer   A printer that can accept output from the computer and print it in full color. Generally, color printers are based on INK-JET or DOT-MATRIX technology.

column Data arranged vertically. For example, spreadsheets are arranged in horizontal rows and vertical columns.

COM In the DOS operating system, an abbreviation of COMMUNICATIONS. For example, communications ports are referred to as COM1, COM2, and so forth.

COMDEX The acronym for COMMUNICATIONS AND DATA PROCESSING DIVISION, a large trade show sponsored periodically by the computer industry to demonstrate new products.

COM File A term used in DOS to label a small program. Larger programs, such as a word processing program, are often comprised primarily of COM files.

command A direction given to the computer to indicate a specific task to be performed.

command buffer A storage area for recently given commands. In a word processing program, for example, the most recent command is stored in a command buffer so that it can be undone through the "undo" command.

COMMAND.COM   A file used in DOS to store commands that are internal to DOS.

command driven   A software program that works through commands the user must learn. This can be contrasted with programs that are menu driven, where commands are chosen from a menu, as well as those in which commands are issued through clicking on an icon with a mouse.

command key   A key on a computer keyboard used in combination with other keys to perform specific functions.

command language   A language used to issue instructions to an operating system. A DOS command such as FILECOPY is an example of command language.

command line   The line on the computer screen where the cursor is positioned and at which point the next command can be entered.

command line interpreter   A function within a computer's operating system that processes commands entered by the user. When the user enters a command, the command line processor interprets it and initiates the execution process. If the command has been entered incorrectly, the command line interpreter issues an error message.

command processor (See COMMAND LINE INTERPRETER.)

comments Notes added throughout a computer program to explain why certain commands and statements are being used and what they are designed to accomplish. Comments help to explain the program to other programmers who, at a later time, might be responsible for making modifications. Comments in a program are labeled as such, so that the computer does not attempt to process them as if they were standard program code.

common carrier A company that provides telecommunications services to the public, such as a telephone company. Common carriers are strictly regulated by the federal government.

communications The process of exchanging information, including the passing of information between devices such as computers.

communications channel The physical means by which communications occur between two computer devices.

communications control unit A device attached to a mainframe computer that serves to monitor and control the telecommunications function.

communications link The method computer devices use to communicate.

communications program A computer program designed to facilitate communications between computers through features such as storing the telephone numbers of communications lines, dialing the numbers, and establishing the communications links.

communications protocol A set of rules for data communications between two or more computers. Without these commonly agreed-upon rules, data would be lost and errors in transmission would go undetected.

communications server Often a small computer, such as a powerful personal computer, that links a local area network to a mainframe or large network. The communications server essentially serves as a go-between, receiving data and passing them on to the personal computers on the local area network, as well as taking data from the local area network and passing them back.

communications software Software programs that are designed to handle the communication function. For example, modems are used in conjunction with communications software that assists the user with functions such as dialing other computers and monitoring the process of making the connection.

compact disk A metal disk on which information can be stored, and later read through the use of a COMPACT DISK PLAYER. Compact disks are also referred to as optical disks, and are abbreviated CD.

compact disk player A device that uses a scanning mechanism, based on laser technology, to read information from a compact disk.

compatibility The ability of one computer to emulate the operations of another, such as run the software designed for another computer. For example, lower-cost IBM clones (see CLONE) are designed with IBM personal computers as the model for compatibility. Compatibility is also possible in the software arena. For example, the ability to use a spreadsheet designed under one spreadsheet program with another is the result of compatibility.

compatible software Software designed to allow for the exchange of data with other software. (See also COMPATIBILITY.)

compile A process in which the program code developed by programmers—the source code—is "translated" into code that the computer can actually understand—the object code. This is accomplished by a program called a compiler. For example, program code such as that developed under languages like COBOL must go through a compiler to be understood by the computer.

compiler The program that performs the compile function.

complementary metal oxide semiconductor (See CMOS.)

complex instruction set computing (See CISC.)

compression The process of packing data into lesser amounts of storage space. Compression is achieved either within the software program, such as a spreadsheet, or through special programs that perform this function.

CompuServe A widely used online information service offering a broad range of information for the business and home, online discussion groups, and other services.

compute   The actions performed by a computer, that is, to store data and perform calculations and other operations, based on commands from users.

compute-bound   A software program that has been limited in speed or functionality due to the size of the computer's central processing unit. For example, running a program on a 286-based personal computer, when it has been designed for optimal use with one that is 386-based, will result in the program being compute-bound.

computer   A machine designed to store and process data.

computer-aided design   (See CAD.)

computer aided design/computer-aided manufacturing   (See CAD/CAM.)

computer-aided engineering   (See CAE.)

computer-aided instruction   (See CAI.)

computer-aided manufacturing   (See CAM.)

computer-aided testing   (See CAT.)

Computer and Business Equipment Manufacturers Association   (See CBEMA.)

computer architecture   The interior and exterior physical design of the computer.

computer art   The use of the computer in the process of producing works of art. The computer artist uses the computer screen as a palette, and an input device such as a mouse to create images.

computer-assisted diagnosis   The use of computers and specialized application programs that assist medical personnel in testing and diagnosing illnesses.

computer-assisted instruction   (See CAI.)

computer-assisted teaching   (See CAT.)

computer-augmented learning   (See CAL.)

computer-based learning   (See CBL.)

computer-based training   (See CBT.)

computer center   The area in a building in which the computers are located. Generally, the computer center houses mainframe and midrange computers and is designed with special air conditioning, raised floors, and a security system.

computer control console   Another term for COMPUTER CONSOLE.

computer crime   The use of computer equipment for criminal purposes, such as illegally tapping into a database with a credit card or other protected information.

computer-dependent   Computer programs that are designed to run on a specific brand of computer, such as IBM or Apple.

computer engineering   The field of engineering focused on the design of computer hardware.

computer family   A line of computers designed and produced by one specific manufacturer. For example, IBM's line of computers can be called a computer family.

computer game   Any game that can be played on a computer, though generally used to refer to arcade-type games that require a joystick. Computer games are characterized by extensive color graphics and, often, sound.

computer graphics Broadly used, the creation of various images, such as charts and drawings, on a display screen, achieved through computer software.

Computer Graphics Metafile A standard file format for computer graphics adopted by a wide range of software programs, enabling the exchange of graphics files.

computer-independent language A language any computer can use providing the necessary compiler program is available. Languages such as COBOL and BASIC are computer-independent.

computer instruction A line of code in a computer program that orders the computer to perform a specific operation.

computer-integrated manufacturing (See CIM.)

computerization The process of adopting a function previously performed manually for use with the computer. For example, word processing, once performed by hand or with the typewriter, is now associated with the computer.

computerized mail Another term for ELECTRONIC MAIL, which is more commonly used. (See definition.)

computer language A language, with its own words and syntax, used to write computer programs. A more widely used term is PROGRAMMING LANGUAGE.

computer literacy Knowledge of computers and how they are used. An individual with a basic level of computer literacy can most likely use one or more basic software programs, such as word processing or spreadsheet. People with a higher level of computer literacy can perform tasks such as setting up a complex computer system, or designing and writing a computer program.

computer-managed instruction (See CMI.)

computer music Music composed and/or performed using a computer.

computer network Linking together two or more computers for the purpose of sharing data and applications.

computer operations The day-to-day maintenance function of the computer, including data backup and security. Large organizations, with mainframe computers, employ a staff of people to perform the computer operations function, often on a 24-hour basis.

computer operator An individual employed in computer operations.

computer program A set of instructions, written in a computer language, that guides the computer in performing a specific task.

computer science Broadly used, the study of computer hardware, software, and data communications.

computer security Broadly used, describes techniques employed to protect valuable hardware and software resources from access by unauthorized users. These techniques include special locks on doors to rooms where computer equipment is stored, and passwords to restrict access to programs.

computer simulation Using the computer to simulate real-life events and situations. For example, specialized software can simulate events such as wars or other catastrophies, as an aid to developing contingency plans.

computer system Broadly used, refers to the combination of the computer processor (CPU), the monitor, and attached devices such as a printer and modem.

concatenate   To link together two or more words into one word. The computer industry is notorious for concatenation, resulting in words such as DATABASE.

concurrent   Occurring at the same time.

concurrent processing   The ability of a computer to perform multiple tasks at the same time.

concurrent program execution   Running multiple programs at the same time.

conductor   A substance, such as metal, capable of conducting electricity.

CONFIG.SYS   The DOS file for directing the way in which the operating system will be set up on the computer. For example, the buffer sizes are designed in this file.

configuration   The setup of a computer system. For example, the number of computers, where they are located, the attachment of printers and modems—all of these are elements in a computer system configuration. Configuration also applies to how a software program will be used, with specifications like the amount of storage space available for program files, and what types of printers will be employed.

configuration file   The file in a software program where configuration information appears.

configure   The process of configuring a system or software program.

connect charge   The fee that a user must pay for the privilege of having access to an online information service or database. Generally, a connect charge is based on an hourly rate.

connecting cable   A cable used with a device such as a printer, or multiple computers with each other.

connectivity   A general computer industry term used to refer to the linking together of computers on a network. Connectivity can refer to both the hardware and software considerations of connecting computers together.

connector   A device to connect two cables together.

connect time   The amount of time a user is involved in accessing an online information service or database. The CONNECT CHARGE (see definition) is based on the connect time.

console   A terminal connected to mainframe computer, for monitoring and directing the operations of the computer. The console is used by the console operator, a member of the computer operations staff.

constant   In programming terms, a value that remains the same at any stage of program execution.

contention   The result of users and programs competing for the limited resources of the computer. If users are competing for access to the same data, the device on which the data is stored may become temporarily unavailable to all, due to contention.

context-sensitive help   Guidance on using a program provided to the user based on the function being performed. For example, if a word processing program offers context-sensitive help, a user having difficulty with a specific step in merging a letter with a list of addresses would, by pressing a key, receive instructions on that step rather than on the whole merge function.

context switching   The ability to switch from one program to another without losing one's place in the first program—for example, partially completing a document in a word processing program, moving to a graphics application for a moment, and then returning to the same spot in the first document.

continuous forms Perforated sheets, usually on a roll, that can be printed one after another without having to be loaded into the printer one at a time. Generally, continuous forms have edges with holes that fit on the printer's tractor feed. Continuous forms can include blank sheets, for documents, or business forms such as checks or invoices.

continuous processing Entering data and executing transactions one after the other, as they occur.

contouring The creation of a graphic image that is filled with shading in ridges and bumps to give the image an illusion of texture.

control The ability to oversee the operations of a computer system so that data are kept safe and reliable, and operations are performed in an orderly manner. Various software products, also referred to as SYSTEM SOFTWARE, are available to facilitate control.

control break A pause in a computer program during which a special processing action is taken, such as printing or performing calculations.

When this action is completed, the program resumes execution.

Control-Break Simultaneously pressing the Control and Break keys in order to issue the BREAK COMMAND.

control character A key pressed in conjunction with the Control key to perform a specific function. Word processing programs often make use of the Control key in conjunction with other keys to, for example, start a new page.

Control key A key on a computer keyboard that, when used with a control character, causes a specific action to be taken.

controller A device in a computer to transfer data to, and control the operations of, other devices like modems and printers. A controller used with a mainframe computer may be the size of a photocopying machine, whereas a controller in a personal computer is most likely contained on a microchip.

control panel The area, particularly on a mainframe computer, where buttons and other devices that initiate manual controls, such as system start-up, are located.

control program A part of the operating system that manages the overall resources of the computer. The control program, for example, ensures that various software programs can all be installed on the computer and will be able to effectively share resources such as memory and storage space.

control statement A statement in a program that stops one operation and initiates another.

control words Words reserved for special use in a software program. For example, the word FILE may be a reserved word in a word processing program, and because it is used for a special purpose, it cannot be used as a document name.

convention A procedure accepted by programmers and other computer professionals. Many of the techniques in programming are conventions that have developed through the years and help to add a layer of standardization among individual approaches to programming.

conventional memory The piece of the computer's memory actually available for programs. A portion of the computer's memory is reserved for use by the operating system.

convergence Combining technologies, such as factory operations and computer operations in the case of CAD/CAM.

conversational Software designed to engage the user in a conversation mode, with the user entering a command and receiving a response which leads to the next command.

conversational interaction Dialogue between the user and a software program.

conversational language A computer language based on relatively simple commands used in everyday human language. DOS commands such as FILE and COPY are examples of conversation language. BASIC also makes use of conversational language.

Conversational Monitor System A relatively easy-to-use, command-based user interface for IBM's VM operating system, which is used on mainframe computers. Abbreviated CMS.

conversion Changing from one computer system to another. Moving from a large mainframe to a network of personal computers is an example of conversion. Another is changing from one word processing software product to another. Conversion often requires extensive planning and careful execution to avoid loss of data and productivity.

conversion table A list of codes and their equivalents, such as standard alphanumeric characters and their binary or ASCII equivalents. (See ASCII and BINARY.)

convert To change from one computer system, or software product, to another.

cookbook A user manual for a software program that includes step-by-step explanations for performing specific functions.

cooperative processing The ability to distribute computer resources, including data and applications, throughout an organization. A computer network of some type is required for cooperative processing.

coprocessor A computer that works with another one to speed up operations, add functionality, or protect computer operations in case the first computer becomes inoperable.

copy To duplicate a set of data. For example, information that is stored on a hard disk may also be copied onto a floppy disk and stored as a backup copy.

copy program A program that provides a simplified means of copying other programs from one storage medium to another. Often, copy programs are used to illegally copy software programs that are otherwise protected from unauthorized copying.

copy protection A practice among some software vendors to assure that their software programs are not copied by an unscrupulous user and passed on to other users to avoid paying additional fees. Copy protection often involves the use of additional files in the program that prevent the program from being copied.

coresident Used to describe two programs loaded in main memory at the same time.

corrective maintenance The process of correcting problems in hardware or software after the problems have become apparent to users. This can be contrasted with PREVENTIVE MAINTENANCE.

correspondence quality Another term for PRINT QUALITY.

corrupt Data that have been changed or damaged in a way that renders them unusable.

cost/benefit analysis A step in the process of purchasing a new computer system, or software program, in which the potential benefit of the system or software is weighed against the potential costs.

cottage workers Employees who work out of their own homes, generally on a personal computer connected to their employer through a modem and telephone line.

country-specific Special symbols and conventions that are specific to a particular country, such as accent marks and ways of denoting date and time. These considerations must be reflected in the selection of hardware and software products.

coupling A connection between computer systems.

courier font A standard font used in basic documents and available on virtually any printer.

courseware Generally, computer-based courses used in an educational setting.

cpi The abbreviation of CHARACTERS PER INCH.

CP/M The abbreviation of CONTROL PROGRAM FOR MICROCOMPUTERS, an operating system for use on personal computers.

cps The abbreviation of CHARACTERS PER SECOND.

CPU The abbreviation of CENTRAL PROCESSING UNIT.

CPU-bound Another term for COMPUTE-BOUND.

CPU cycle The computer's basic unit of time, generally a fraction of a second. An operation is measured in CPU cycles.

CPU time The amount of time required by the CPU to execute a given task. CPU time can be affected by factors such as the numbers of users making demands at the same time.

crash The complete failure of a computer, resulting from an event such as a serious program error or a power outage.

create To build something new, such as a new document, as opposed to modifying one that already exists.

critical error An error severe enough that the operation of the computer is halted until the user corrects the condition. For example, sending a document to a printer that is not in operation will result in a critical error.

critical path   The shortest distance between the points on a network.

crop   To cut a piece from a picture or a graphic image.

cross-check   To verify the results of an operation, such as a database search or a calculation, by performing the operation in two different ways and comparing the results of both.

cross-compiler   A compiler that works on more than one computer.

crosstalk   Garbled signals over a network, resulting in static and unusable data.

CRT   The abbreviation of CATHODE RAY TUBE, which is the technology used to create images on the screens of televisions and most computer monitors.

crunch   A computer industry slang term for the computer's ability to handle large amounts of numerical data and perform arithmetic operations.

Ctrl   The abbreviation of CONTROL KEY.

cursor A symbol on the computer screen indicating where the next character can be placed. Generally, the cursor is a blinking dash.

cursor control keys Keys that direct the movement of the cursor. Also referred to as arrow keys, because each is labeled with an arrow.

cursor position The location of the cursor on the screen. Many word processing programs indicate the current location of the cursor to aid the user in performing functions such as creating columns.

customer support The value-added assistance provided by hardware and software vendors. Customer support may include a special telephone number, or hot line, that registered users can call to receive special assistance from a trained customer support technician.

customize To change or modify a hardware system or a software product so that it better meets the user's needs. For example, attaching a mouse to a personal computer is a means of customizing the system. And software products often include special features that enable the user to easily modify the product for his or her own situation.

custom software A computer program developed expressly to meet the needs of one specific user, department, or organization. An example of custom software is a financial accounting system developed for a specific user.

cut In word processing, removing part of a document, such as a paragraph, and either moving it to another location in the document or discarding it altogether.

cut-and-paste Moving part of a document from one location to another.

cut-sheet feeder A mechanism that feeds sheets of paper into a printer, such as that built into a laser printer.

cylinder A column of tracks on a hard disk, used to divide the disk into segments.

D

DA 1. The abbreviation of DESK ACCESSORY, a program available on both Apple Macintosh and graphical user interface programs for IBM and compatible personal computers. A desk accessory is a computer-based version of some of the things a user might have on his or her desktop, such as a calendar and a clock. 2. The abbreviation of DIRECT ACCESS, meaning that data can be accessed directly, without having to scan through other data during the process. This is also referred to as RANDOM ACCESS.

daisy-wheel printer An outdated impacting printing technology in which the characters are embossed on a metal wheel. Printing occurs when a hammer strikes the wheel against the ribbon.

While the print quality with a daisy-wheel printer is high, the process is slower than with other printing technologies like dot matrix.

DASD The abbreviation of DIRECT ACCESS STORAGE DEVICE, a storage medium, generally used with mainframe and midrange computers that allows data to be positioned so that they can be accessed directly and pulled into main memory, without having to scan through other data during the process.

DAT The abbreviation of DIGITAL AUDIO TAPE, a medium for data storage. A DAT is a tape contained in a cartridge device about the size of a credit card.

data Information that has been formatted to be stored and processed by the computer. This can include numbers, text, and program code.

data attribute Information that describes the type of data contained in a field on a database record. For example, the attributes of a field containing the social security number include a field length of nine numeric characters.

data bank A repository containing data, often organized around a specific topic or topics.

database   An electronic "file cabinet" containing data that are somehow related. For example, the personnel database might contain the records of all employees in an organization, while the sales database might contain records of all recent sales transactions.

database administrator   An employee in the information services department who is responsible for maintaining the database management system, including issuing standards for the abbreviation and formatting of data.

database management system   A computer program developed to manage the development and use of database. Database management systems include facilities for building databases and formatting and retrieving records. This might include easy-to-use menus to assist users in this process.

database manager   Another term for DATABASE MANAGEMENT SYSTEM.

database server   A computer attached to a network that has the sole task of storing data. The database server responds to requests for data from other users on the network.

data buffer   Another term for BUFFER.

data capture Generally used in reference to systems that include computers at geographically dispersed sites that feed information to a central computer. For example, cash registers can collect point-of-sale information that is fed to a central computer.

data collection The process of entering data into a computer system.

datacom A slang term for DATA COMMUNICATIONS.

data communications The area of computer science that deals with coding and sending data by electronic means, including both hardware and software considerations.

data compression Making storage more efficient through techniques that reduce the amount of space required to store data. This might include removing spaces between words. Spreadsheets and graphics programs often include data compression options.

data dictionary A file, or list, that serves as a directory to all of the data stored in a database management system—all databases, a list of the fields in each, and the size of each. The data dictionary is maintained by the database administrator, and users do not have access to it.

data element   A piece of information included on a data record. For example, the current date and number of items sold are data elements included on a sales record.

data encryption   Protecting sensitive data by converting them into codes that prevent them from being read by general, unauthorized users. Data encryption is achieved through special software programs designed for this purpose.

Data Encryption Standard   A standard for DATA ENCRYPTION developed by the U.S. National Bureau of Standards.

data entry   The process of adding data to a database, generally by typing them in on a keyboard.

data entry device   A mechanism, such as a keyboard or a scanner, for entering data into a computer.

data export   The ability to create information in one software package that is also usable in another. Many word processing programs include this feature, with the option of formatting documents in a standard format, such as ASCII, that other programs can use.

data format   The means by which data are organized, or structured, in an application program or a database management system.

data import   The ability to "pull in" information developed under one software program for use in another. An example of this process is writing a document under one word processing program and subsequently editing it in another.

data integrity   The overall trustworthiness of data, assuming that they have been entered correctly and checked for accuracy.

data manipulation   The process of using, or changing, data. Examples of data manipulation include creating a spreadsheet or retrieving data from a database.

data network   Broadly used to refer to computers that are somehow connected to each other for the purposes of sharing data.

data point   A numeric value on a graph or chart.

data processing   A general, and somewhat outdated, term referring to the functions associated with storing and processing data. The more accepted terms for these functions include INFORMATION MANAGEMENT and INFORMATION SCIENCE.

data processing center The physical location where data are entered and stored into the computer for subsequent retrieval by users.

data security Protecting data from being read and manipulated by unauthorized users, or accidentally destroyed. Data encryption is one method of enhancing data security. Requiring users to have passwords is another.

data sharing The use of one source of data, such as a single database, by multiple users. An example of data sharing is sending data files from one computer to another through telecommunications lines.

data structure The organization of data in structures such as files or records. Data structure is based on the way data are being maintained and retreived. For example, a personnel record is structured to be retrieved by data elements such as name and employee number.

data transfer The movement of data from one place to another. Data can be transferred by means of telecommunications lines, over a modem, for example, or across a network, as in a local area network of personal computers.

data type In programming, an identifier given to data to direct the program in how to use them. For example, text data are identified as such in a computer program, to distinguish them from numeric data.

date The current month, date, and year, based on the computer's internal clock. If the user has not taken the time to set the clock, this information will be displayed as a series of zeroes.

DB The abbreviation of DATABASE.

dBase A popular, personal computer-based, database management program.

dbms The abbreviation of DATABASE MANAGEMENT SYSTEM.

DC The abbreviation of DIRECT CURRENT, the power provided through an electric outlet. This current flows in only one direction.

deallocate To free up an area of main memory that was previously allocated to a program or data.

debug To correct errors in software.

debugger A program to locate bugs in other software programs.

DEC   The abbreviation of DIGITAL EQUIPMENT COR-
PORATION.

decentralized processing   The dispersal of computer resources (from microcomputers to mainframes) to one or more geographical locations,
with computer processing tasks shared among
these locations. For example, an organization
with multiple branch offices might locate a computer in each office. Computers in a decentralized processing environment may or may not be
connected over a network, though they generally are connected.

decision   When a computer program evaluates a
piece of data and subsequently takes an action
based on that variable. For example, in a check
balancing program, a negative number would
result in the program issuing a overdraw notice.

decision box   A symbol in a FLOWCHART to indicate
a point in a computer program at which a decision is made. A decision box is shaped like a
diamond.

decision support system   A computer program
that aids in analyzing information and making
decisions. Decision support systems are designed to guide the user in listing alternatives,

evaluating each one, and making an informed decision.

decision tree A diagram that shows the points in a program at which decisions are made, and the alternate directions that can be taken.

decoder A computer program to restore coded data to their original state. An electronic device that performs this function can also be referred to as a decoder.

decrement The process of performing subtractions.

decryption The process of restoring encrypted data to its original state. (See also ENCRYPTION.)

dedicated Designated for one purpose only.

dedicated device A device such as a printer that has been designated for one purpose. For example, a printer used only for printing checks, and no other purpose, is a dedicated device.

de facto standard A method, or program, used so often, by so many individuals, that it becomes an accepted standard without having been formally designated as such. For example, the DOS operating system is in such common use that it has become the de facto standard operating system for IBM and compatible personal computers.

default   The choice made by a computer program in the absence of any other being specified. For example, most word processing programs assume single line spacing if the user does not indicate an alternative.

default drive   The disk drive to which the operating system first attempts access, unless the user indicates otherwise. For example, during the startup process, DOS generally "looks" first in the floppy, or "A" drive, for an application program or a user interface. (See also START-UP.)

deinstall   To remove a software program from a computer.

delete   To erase.

Delete key   The key on the standard keyboard that, when pressed, erases the character above the cursor.

delimiter   A special character, like a back-slash or quotation marks, for the purpose of separating one string of characters from another.

Del key   (See also DELETE and DEL KEY.)

demonstration program A sample of a larger program offered, on a diskette, for potential customers. Demonstration programs contain examples of major program functions, often requiring some interaction by the user, accompanied by graphics and even animation.

demo program (See DEMONSTRATION PROGRAM.)

density The amount of data that can be physically stored on a storage medium. For example, high-density diskettes can hold more data than low-density diskettes.

desk accessory In software programs, utilities like a calculator that are always available to the user, regardless of the application that is currently being run. (See DA.)

desktop computer A term used for a PERSONAL COMPUTER, because it is self-contained unit designed to fit on an average desk.

desktop publishing Using a personal computer and high-quality laser printer to design and print documents that previously would have required the services of a professional. Desktop publishing requires special software programs designed to guide the user through formatting the document into columns, placing headers over the articles, and including graphics. While desktop publishing can be done with only a basic personal computer, a simple desktop publishing program, and small laser printer, at minimal cost, professionals use high-end personal computers and workstations powered by sophisticated desktop publishing software and fast, full-featured laser printers.

destination   A term in data communications to indicate where the data are being sent. The destination can be another user's computer, for example, or a database.

device   Any equipment that can be attached to a computer.

device dependent   Programs that can only be run on certain hardware, such as those developed for use with the Macintosh.

device driver   A computer software program to run a certain device. For example, a word processing program requires the installation of a special driver for each printer attached to it.

dialog   The process of interacting with the computer by answering questions posed by the computer program.

dialog box Boxes that appear on the computer screen to request information from the user before a specific function can be performed. Generally, once the user enters the information, the dialog box disappears. Graphical user interfaces and easy-to-use programs often employ the use of dialog boxes.

digit A character in the numbering system.

digital The use of numbers only—0 and 1—as the basis of coding a computer.

digital audio tape (See DAT.)

digital data transmission The sending of data in the form of bits, which normally occurs when data flow from one computer device to another.

Digital Equipment Corporation A large manufacturer of computer hardware and software, including the VAX line of computers, as well as workstations and personal computers.

digital video interactive Technology that enables a computer to act like a television, storing video images and displaying them fast enough to indicate movement.

digitize To convert a picture or drawing into digital code so that it can be reproduced on a computer screen.

digitizer A device used in the process of digitizing.

digitizing tablet The most common digitizer, consisting of a flat panel with a grid of wires underneath. The user lays the graphic on this panel and traces over it with a pen or stylus attached to the computer. The diagram is then converted into code and stored in the computer.

dimensioning A process in design programs to indicate the physical dimensions, such as height and length, of an image. This is often achieved by including information such as the proposed measurements of the frame.

dingbat Small graphics such as stars and snow-flakes, provided with graphics packages, that can be imported for use in a document. This is a simple way of illustrating a document without having to create graphics from scratch.

DIP The abbreviation of DUAL IN-LINE PACKAGES, a device contained on printers and other computer equipment to set various controls that allow the equipment to work with specific software programs. The DIP consists of a row of switches mounted on a circuit board.

DIP switch   A tiny lever used on a DIP. The row of DIP switches are set in on–off positions.

direct access   (See DA.)

direct access storage device   (See DASD.)

direct-connect modem   A modem that plugs directly into a telephone jack.

direct current   (See DC.)

direction key   Another term for ARROW KEY, used to move the cursor to different positions on the display screen.

directive   Another term for COMMAND.

directory   A guide to the files that are contained in storage. Directories are hierarchical and can be thought of in terms of a tree structure. For example, the hard disk directory contains information about all of the software programs that are currently stored on the hard disk. And the directory within each software program lists the files, such as word processing documents or spreadsheets, that have been created and stored within that program. Additionally, users can create subdirectories of related files.

disable To intentionally prevent a function, in a hardware device or computer program, from being used. For example, a computer's modem can be disabled if it is not needed for a period of time, or a program can be disabled if it is not needed.

disc An alternate spelling for DISK, generally in reference to OPTICAL DISCS rather than FLOPPY DISKS.

discrete Separate or distinct, as in discrete items of data.

disk A device on which data can be stored. A disk is round in shape, but held in a square plastic casing. Magnetic storage disks can be either hard if a hard or fixed disk, or flexible if a floppy disk. Optical storage disks are made of a rigid material.

disk access time The time required to locate and retrieve information from a disk.

disk cache A part of the hard disk used for CACHE MEMORY.

disk crash   A disk malfunction that destroys all the data stored on the disk or otherwise renders it unusable. Also called HEAD CRASH.

disk drive   The mechanism in the computer that transfers data from memory and writes it onto a disk, and that reads data from the disk and transfers it into memory. To do this, the disk drive rotates the disk like the record on a turntable. Different types of disk drives include hard, floppy, and CD-ROM.

disk envelope   A paper or plastic covering for a diskette that protects it from being contaminated by dust or from being scratched.

diskette   Another term for a FLOPPY DISK of any size.

disk jacket   Another term for DISK ENVELOPE.

diskless workstation   A personal computer attached to a local area network, or to a midrange or mainframe computer, that has no disk drives. The user has all the capabilities of the personal computer except for the ability to store data. Instead, data are retrieved from, and subsequently stored on, a computer that acts as the network file server.

disk operating system   The most commonly used operating systems for IBM and compatible personal computers, developed by the Microsoft Corporation. Generally referred to by its acronym, DOS.

disk optimizer   A software program that rearranges the data stored on the disk to make both storage and retrieval more efficient. For example, the program eliminates large gaps between pieces of data.

disk pack   A group of hard disks stacked and encased in a protective container. A disk pack is treated as one disk by the computer.

disk unit   Another term for DISK DRIVE.

display   To represent data on a screen; also a slang term for DISPLAY SCREEN.

display adapter   An expansion board that connects the computer with a monitor. It acts both to determine the capabilities of the monitor, as well as to enhance these capabilities.

display card   Another term for VIDEO ADAPTER.

display device   Another term for DISPLAY SCREEN or MONITOR.

display screen The part of the monitor that displays data.

distributed computing system A network of computers that work independently of one another. For example, each department in the organization may have its own computer or network of personal computers, and be responsible for maintaining its own set of departmental data, while messages to and from other departments are shared over the network.

distributed database A database that is physically located in more than one location of the organization but connected by the computer network. Distributed databases allow users access to the data they use most often in their work. Distributed databases are managed by a distributed database management system.

distributed processing An approach to information management in which data are stored and processed on more than one computer.

dithering Mixing together various color dots on the computer screen to create new colors, similarly to the way in which an artist mixes colors on the palette.

document   A file containing text, generally created with a word processing program.

documentation   The printed manual, or set of manuals, that accompanies a hardware or software product. The documentation is a guide to the installation and use of the product.

document file   A file containing a document or other data created by a user as a result of working with an application program. For example, a document created through a word processing program is stored in a document file, as opposed to the program file in which the actual program is stored.

document processing   Broadly used, the creation, ongoing editing, and storage of documents. Document processing is the overall function of a word processing program.

document retrieval   The process of searching for stored documents based on unique search criteria—for example, searching all documents in storage for those that contain a reference to a certain event, such as "Civil War," or a certain date, such as "1865." Document retrieval requires the use of sophisticated software products designed for use with mainframe, midrange, and personal computers.

DOS   The abbreviation of DISK OPERATING SYSTEM.

DOS prompt   The message on the screen indicating that DOS is ready for a command. Generally, the DOS prompt is composed of the letter identifying the drive being used followed by the "greater than" symbol. (See also DRIVE LETTER.)

dot   Another term for PIXEL, a tiny point or speck on a display screen that is combined with others to form graphic images. Screen resolution is expressed in DOTS PER INCH.

dot-matrix printer   Printer technology based on creating letters composed of closely spaced dots. A small hammer strikes a mechanism with pins that in turn hit the printer ribbon. Dot-matrix printers are realtively inexpensive and fast. Those attached to personal computers are generally 24-pin, which offer a high level of letter quality.

dot pitch   A measurement used to indicate the quality of resolution on a display screen. Dot pitch is measured in millimeters; the closer together the dots are, the greater the resolution. Color monitors begin at a dot pitch of 0.22 mm.

dots per inch   A measurement used to indicate the quality of type provided by a laser printer. The more dots per inch, the greater the resolution.

Laser printers generally offer a dots per inch of 300 or above. The abbreviation for dots per inch is DPI.

double click The process of invoking a command by pressing a mouse button twice in succession.

double-density disk A floppy disk with twice the storage capacity as a single-density disk, in the same amount of space. Double-density, 5¼-inch disks can hold 360K of data, while double-density, 3½-inch disks can hold 720K.

double-sided disk A floppy, or magnetic, disk that allows for the storage of data on both sides.

down A computer system that is currently out of use due to an event such as a power outage, or because it has simply been turned off.

download To transfer data from one device to another. The term generally implies that data have been transferred from a larger system, such as a mainframe computer, to a personal computer or workstation.

downloadable font A font that has been transferred from the computer's memory to the printer for use on a specific task. By contrast, RESIDENT FONTS are stored on the printer.

downtime A period of time during which a hardware device, particularly a computer, is nonfunctional. The average amount of downtime experienced by a device is a way of measuring its reliability.

downward compatible New versions of hardware or software products that can be used with older versions of the same product. For example, a new version of a word processing program is downward compatible if it can be used to modify documents that were created with older versions of the program.

dp The abbreviation of DATA PROCESSING.

dpi The abbreviation of DOTS PER INCH.

draft mode A much faster, yet lower-quality print output available on some printers. Dot-matrix printers can often be used in a faster, lower-resolution draft mode, or in a slower, but higher resolution NEAR-LETTER-QUALITY mode.

draft quality The quality of print that results from using a printer in draft-quality mode. Generally, the print is lower in resolution and lighter.

drag The process of moving an icon from one location to another by using the mouse to click on the icon, holding the button down as the icon is moved, and then releasing the button.

draw program A graphics program used to create images on the computer screen. The program may include features such as templates to assist in creating shapes and images, and offer options for filling in the images with color. These drawings can then be imported into a document or stored as images.

drive A slang term for DISK DRIVE.

drive bay A space in the cabinet of a personal computer for installing a disk drive. Generally, personal computers have drive bays for two disk drives—one for 5¼-inch floppy disks, and one for 3½-inch floppy disks.

drive letter The alphabetical means by which all storage device drives, including floppy disk, hard disk, tape, and optical, are identified by the operating system and application software on IBM and compatible personal computers. These begin with the letter A, followed by a colon. For example, the first floppy disk drive is identified by "A:", while the hard disk drive is identified by "C:".

drive number   The means by which all storage device drives are identified on Apple Macintosh personal computers. The drives are identified by numbers, beginning with 0.

driver   A set of instructions used by the computer to communicate with a specific peripheral device, such as a modem or printer. For example, word processing programs require the use of a separate driver for each printer type and model used with the program, to accommodate the features and design of the printer. Drivers are provided on diskettes.

drop-down menu   A temporary menu that appears when a command is selected with a mouse. The items on the drop-down menu support the command that was selected. Choices are made by clicking with the mouse, after which the drop-down menu disappears.

DSS   The abbreviation of DECISION SUPPORT SYSTEM.

DTP   The abbreviation of DESKTOP PUBLISHING.

dual density   The ability of a floppy disk drive to read and write to both low-density (1.2 MB) and high-density (360 KB) disks.

dual disk drive Used in describing a personal computer with two floppy disk drives. Usually, one of these drives is designed to handle 3¼-inch disks, and the other 5¼-inch disks.

dual processors The use of two processors within the same computer, generally for the purpose of increasing speed and efficiency. Personal computers sometimes contain a second processor to add graphics and mathematics capabilities.

dual-sided disk drive A disk drive designed to read and write data on both sides of a disk. Dual-sided disk drives require the use of DUAL-SIDED DISKS.

dumb terminal A computer terminal that has no processing capabilities, such as might be attached to a mainframe computer.

dump Printing, or otherwise displaying, raw data without taking time to format or otherwise organize it. For example, in the case of a system failure, the contents of main memory, or of a database, can be printed in its entirety. These data can then be analyzed for clues as to what went wrong.

duplex printing Printing on both sides of a sheet of paper.

DVI The abbreviation of DIGITAL VIDEO INTERACTIVE.

Dvorak keyboard A keyboard designed during the 1930s by August Dvorak. The keys are arranged in a different order from the traditional QWERTY keyboard, based on the premise that the Dvorak arrangement allows for faster typing.

dynamic Unpredictable, or constantly changing. In computer science, conditions related to both hardware and software products, such as use of main memory, that are constantly in a state of flux.

E

EBCDIC The abbreviation of EXTENDED BINARY-CODED DECIMAL INTERCHANGE CODE, which represents data as numbers. It is used in many larger computers, though personal computers often use ASCII code.

ECMA The abbreviation of EUROPEAN COMPUTER MANUFACTURERS ASSOCIATION, makers of computers and related hardware, based in Geneva.

edge connector The elongated electrical socket on a printed circuit board that is used to plug it into the motherboard of a computer.

EDI The abbreviation of ELECTRONIC DATA INTER-CHANGE, in which companies engage in computer-based order-and-invoice processing through the use of telecommunications lines. EDI is much quicker than traditional order-and-invoice processing through the mail system and eliminates paperwork.

edit To check data for correctness, and to make changes as necessary.

edit mode The condition in which a computer program is ready to be modified, generally by a programmer, rather than ready for a user to enter a command.

editor A computer program used to modify text files.

EDP The abbreviation of ELECTRONIC DATA PROCESSING, another term for data processing, referring to the processing of information with the use of a computer.

EGA The abbreviation of ENHANCED GRAPHICS ADAPTER, a graphics system for personal computers that supports sixteen colors and resolution of 640 by 350.

EIA The abbreviation of ELECTRONIC INDUSTRIES ASSOCIATION, an industry trade association.

8086   A shortened term for the 8086 microprocessor, manufactured by the Intel Corporation. Earlier IBM and compatible personal computers were based on the 8086 microprocessor.

8088   A shortened term for Intel's 8088 microprocessor, which was also used in earlier IBM and compatible personal computers.

EIS   The abbreviation of EXECUTIVE INFORMATION SYSTEM, a computer program that assists executive decision makers in organizing facts and other information for use as input into the strategic planning process. Executive information systems include flexible report-creation facilities.

EISA   The abbreviation of EXTENDED INDUSTRY STANDARD ARCHITECTURE, a bus architecture designed for IBM-compatible 386 and 486 personal computers.

ELD   The abbreviation of ELECTROLUMINESCENT DISPLAY, the technology behind the flat-panel displays used on notebook computers.

electroluminescent display   (See ELD.)

electronic bulletin board   Another term for BULLETIN BOARD SYSTEM, or BBS.

electronic cottage The concept of home offices, equipped with computer equipment and modem connections, that allow workers to be employed at home.

electronic data interchange (See EIS.)

electronic data processing (See EDP.)

electronic mail The ability to communicate by sending and receiving messages from one computer to another, either within the same building, or to locations worldwide. Electronic mail is also referred to as E-MAIL.

electronic music (See COMPUTER MUSIC.)

electronic office Describes a home office that features computer equipment with business programs such as word processing and spreadsheet.

electronic publishing Broadly used, the distribution of information through electronic media, such as CD-ROM, rather than the traditional printed page. Electronic publishing can also refer to the use of desktop publishing software rather than traditional typesetting methods.

elegant Computer code designed with the utmost skill.

element An item, or piece, of data.

elite Generally used to refer to a smaller size of type, printed at 12 characters per inch. Elite type can be contrasted with the larger-sized PICA type, which is printed at 10 characters per inch.

E-Mail (SEE ELECTRONIC MAIL.)

embedded command In word processing, commands that are inserted in the body of text but do not actually print. Page breaks and underline commands are examples of embedded commands.

emulate The ability of one hardware or software program to imitate another. When one product is generally accepted as the industry standard, other products are developed with features that emulate those of the first, in some cases as a lower-cost alternative, as a means of expanding market share.

enable To allow a computer device or software program to operate, through turning on a switch, for example, or entering a command.

encapsulated PostScript The graphics file format used in PostScript, a popular print language used in desktop publishing applications.

encryption Protecting computer code by translating it into another code that makes it impossible to read by unauthorized individuals.

End key A key that, generally, immediately moves the cursor to the end of the line on which it is positioned. The function of the End key can vary among different software programs.

end-of-file The condition in a computer program reached when all the data in a file have been processed. Also, a key that is used to mark the end of a file.

end user Users of computer applications who do not have in-depth technical skills—for example, the individual who actually makes use of a word processing or graphics program, as opposed to the person who develops it, is the end user. Professional accountants who use accounting programs are also referred to as end users.

Enhanced Graphics Adapter (See EGA.)

enhanced keyboard A larger, more fully functional keyboard than that used on earlier personal computers. Used with 386 and above computers, it contains 101/102 keys. A major difference on this keyboard is that the function keys are located in a row at the top of the keyboard, as opposed to being arranged on the left side.

Enter key The key used to actually execute commands; also referred to as the RETURN KEY. Depending on the software program, can also be used for other purposes, such as starting a new paragraph in word processing.

environment A very generic term for the combination of hardware, operating system, and applications. For example, working in a DOS environment is different from working in a large mainframe environment.

EO The abbreviation of ERASABLE OPTICAL DISK, an optical disk that can be erased so that new information can be stored on it. Most optical disks are of the CD-ROM variety, which can only be used for reading data.

EOF (See END-OF-FILE.)

EPS See (ENCAPSULATED POSTSCRIPT.)

erasable optical disk (See EO.)

erasable storage Storage devices that can be used more than once, including hard and floppy disks and magnetic tape.

erase To remove data from a storage medium.

ergonomics The study of man–machine interactions, with a focus on designing equipment that complements natural human abilities and limitations. Chairs that offer adequate back support, and nonglare computer screens, are examples of practical applications of ergonomics.

error A mistake that interferes with the completion of a task. An error can affect either software or hardware. An error can result from a range of factors, including damaged equipment, a program bug, or user misjudgment.

error checking The process of detecting inaccuracies in data being sent over a telecommunications line. Generally, error checking is performed by software designed for this task.

error detection Using hardware and software techniques to detect errors in equipment and programs. Each category of equipment, particularly in the telecommunications area, has its own set of error detection practices.

error file A file that may be generated during a computer session to record any errors that occurred during an operation. Software programs installed on mainframe computers, or local area networks, collect this information every time an error condition is reached. The error file can later be analyzed to better understand the kinds of errors that are occurring, and when.

error handling The process of managing errors that occur while a program is being executed. Some programs have built-in capabilities for self-correction, while others either display a message for the user or terminate.

error message A message provided by the software when an error condition is reached. Generally, a message appears on the screen, either briefly describing the error or identifying it with a number that the user can look up in an error manual.

escape character The character that results from pressing the Escape key.

Escape key (See ESC KEY.)

ESC key The abbreviation of ESCAPE KEY, used to terminate an operation before it is completed, or to exit from a software program.

Ethernet A set of protocols that forms the basis of local area network systems. The Ethernet protocols provide guidelines for the speed and method of transmitting data.

European Computer Manufacturers Association (See ECMA.)

evaluate To take a critical look at a hardware or software product to understand what the vendor promises as opposed to what it actually delivers, often an extensive process undertaken before a purchase is made.

even header A header that only appears on the even-numbered pages of a document.

event Broadly used, a situation or occurrence that takes place during the execution of a computer program. A user pressing the Enter key can be described as an event, as can an error that occurs during processing.

exception Generally, an error severe enough to interrupt a computer program's execution.

executable file A file that contains program code that cannot be understood by humans but can be directly executed by the computer.

execute To obey a command and perform the desired operation, such as might be issued by a user to a computer program.

execute cycle The time it takes the computer to accept a command and actually complete the required operation.

execution time (See EXECUTE CYCLE.)

executive information system (See EIS.)

EXE File An executable file under the DOS operating system. It is identified with an EXE extension.

exit To leave a program.

Expanded Binary-Coded Decimal Interchange Code (See EBCDIC.)

expanded memory The ability to add to the capacity of a computer's memory through hardware or software techniques.

expansion board An electronic card (see CIRCUIT BOARD) inserted into the computer system to provide additional capabilities, such as graphics or communications.

expansion card Another term for EXPANSION
BOARD.

expansion slot A slot in the computer for the insertion of an expansion board.

expert system A computer program that emulates
the decision process undertaken by humans and
is thus able to supplement, and even substitute
for, human intervention. Programs with expert
system capabilities are designed to analyze a
range of factors, simulate a situation, make decisions about the best action to take, and actually perform the operation based on this
information. An example of a program with expert system capabilities is a climate control system in a building, which analyzes factors like
temperature and time of day, and control the
climate accordingly.

exploded view A small section of an image that is
being shown enlarged. For example, an engineer
using a CAD system to design an automobile
might display a much larger screen image of one
of the headlights.

export To produce data that can be used by, or
imported into, other programs.

expression Any number or combination of numbers, such as 4 or 4 + 2.

extended ASCII Codes that extend the basic ASCII codes to include additional symbols, such as those in foreign languages.

extended graphics array A high-quality standard for color graphics on IBM and compatible personal computers. Extended graphics array supports up to 256 colors with 1,024-by-768 resolution. The abbreviation is XGA.

Extended Industry Standard Architecture (See EISA.)

extended VGA Enhanced VGA that offers greater resolution, either 800 by 600 or 1,024 by 768. Also referred to as SUPER VGA.

extension An additional feature beyond what was originally installed on a computer system.

external command A DOS term referring to commands that are not included in the COMMAND.COM file.

external modem   A modem contained in a case that is separate from the computer, rather than installed internally on an expansion board.

external storage   Storage devices external to the computer and attached with a cable. A separate hard disk or tape drive is an example of external storage, which is also referred to as AUXILIARY STORAGE.

F

f The abbreviation of FREQUENCY.

facilities A catch-all term for computer equipment. Generally associated with large, mainframe computers and related peripheral devices, such as printers and communications controllers.

facsimile machine A device that scans images and text, and sends them over a telephone line to a compatible device. Most facsimiles use thermal paper though, increasingly, standard cut-sheet paper is also being used.

fail-safe system Another term for FAULT-TOLERANT SYSTEM.

failure   The inability of a hardware device to function, through physical occurrences like a power outage, mechanical problems, or errors in installation.

fan   A cooling device installed in the cabinet of a personal computer or a printer to prevent overheating.

fanfold paper   Paper designed to be continuously fed into a printer, with each sheet separated by a perforated line and holes along each side that fit on the pins of a tractor feed. Fanfold paper is generally used with dot-matrix printers. (See FEED and TRACTOR FEED.)

fatal error   An error that occurs during the execution of a program; as a result, the program is unable to continue.

fault-tolerant system   A computer system designed to perform with a level of reliability such that, even in the event of a hardware or software error, it functions as expected. This may be accomplished by a range of design innovations, including two processors that work in conjunction with each other.

fax   The acronym for FACSIMILE.

fax board   An internal modem that enables documents to be sent directly from the computer (without being printed) to either another fax modem, or to a standard facsimile machine. Internal modems often include facsimile capabilities.

fax machine   The acronym for FACSIMILE MACHINE.

fax modem   Another term for FAX BOARD.

FCC   The abbreviation of FEDERAL COMMUNICATIONS COMMISSION, the federal agency responsible for regulating communications as well as dictating standards for equipment such as personal computers.

FDDI   The abbreviation of FIBER-OPTIC DIGITAL DEVICE INTERFACE, a set of protocols for sending data, voice, and video over fiber-optic lines.

feathering   In word processing, adding spaces between words on a line so that all lines are right-justified.

feature   A particular task that a computer, or software product, will perform. For example, the ability to print in full color is a feature of some printers.

Federal Communications Commission (See FCC.)

feed To advance paper through a printer, as with fanfold paper on a tractor feed. Feed can also refer to the process of sending data to a computer.

Fiber-Optic Digital Device Interface (See FDDI.)

fiber optics A data transmission technology that uses glass or plastic threads to transmit data at the speed of light.

field An area on a database record in which a specific unit of data is contained. For example, personnel records include a field for the employee's name, a field for social security number, and a field for job title.

file Broadly used, a collection of related data identified by a file name. For example, a document created and stored through a word processing program is referred to as a file, as is a spreadsheet.

file allocation table A table, or list, within the operating system that keeps track of a user's files and their locations. The system uses this table as users create and modify files.

file attribute   A characteristic or quality of a file.

file backup   A copy of a file on a separate storage medium, such as a floppy disk, to be available in the event of damage to the original copy.

file compression   Using special techniques to compress the data in a file so that it requires less space on a storage medium. Similar to DATA COMPRESSION.

file conversion   Translating the data contained in a file from one format to another, while the actual contents of the file remain unchanged.

file format   The way in which data in a file are organized and stored. The format of a document file, created in a word processing program, is relatively simple; a file with graphic images is much more complex.

file locking   The process of making one file available to only one user at a time. This is necessary in systems that accommodate multiple users, such as local area networks, so that, if one user is modifying a file, other users cannot access the file at the same time and make additional modifications. Once the user making the modifications is finished, the file is unlocked and made available to other users.

file management The process of organizing the files in a computer system. While the operating system does this, additional file management capabilities are available in each application program, and special file management software programs can also be purchased.

filename The user's name for a file. Generally, a filename can be up to eight characters in length.

file protection Taking precautions to assure that files are not accidentally damaged. Placing tape over the notch on a diskette, to prevent any additional data to be written on it, is an example of file protection.

file security Protecting the data in a file from being read or modified by an unwanted user. Attaching a password to a document stored in a word processing program is an example of file security.

file server A computer on a network that stores files available in turn to the users on the network. Users may also create their own files and store them on the file server, to be made available to other users.

file sharing The ability of many users to access the same data files. Examples of file sharing include different users accessing the same files on a mainframe computer, as well as on a local area network.

file transfer Moving a file from one location on the network to another, such as from one computer to another.

File-Transfer Access and Management A communications standard in transferring data from one computer to another.

file type The nature of the data in a file, usually indicated in the name of the file or in an extension to the file name. For example, the file type of a file stored within the DOS operating system is identified by an extension such as COM to identify it as a communications file.

fill Another term for PAINT.

finder A computer program, associated with the operating system, that organizes the files on a disk. FINDER is also the name given to the file management system on the Apple Macintosh computer.

finite To have specific limits.

firmware A combination of hardware and software, in the form of a program that has been installed on a read-only device and thus embedded in permanent memory.

fixed disk Another term for HARD DISK.

fixed length Generally, a field that always has the same number of characters. The field on a record that contains the social security number is fixed length.

fixed pitch A font in which every character is the same width.

fixed width Another term for FIXED PITCH.

fixed word length Another term for FIXED LENGTH.

flag A mark or symbol that indicates a unique situation, often provided through the program. For example, a list containing potentially incorrect data would contain flags that indicate this possibility.

flatbed scanner A device for performing optical scanning on a flat surface, often used for documents.

flat-panel display A small, very thin screen often used on laptop computers. Flat-panel displays use either gas plasma or liquid-crystal display technologies.

flat screen Another term for FLAT-PANEL DISPLAY.

flat technology monitor A standard-size computer terminal with a flat, rather than slightly curved, screen. The flat design of the screen is used to reduce glare.

flicker A fluctuation that occurs on a computer screen when the display is not refreshed, or withdrawn, rapidly enough. Flicker can also be the result of individual differences in perception.

flight simulator A combination of hardware and software that simulates the experience of flying an aircraft, used in computer games as well as for training pilots.

flippy diskette A slang term for a diskette that stores data on both sides.

floating-point unit A chip that performs floating-point calculations, which greatly enhances the speed of specialized applications such as graphics.

floppy disk A flexible diskette, generally 3½ inches or 5¼ inches in size (though larger and smaller sizes are also available), that is inserted in a floppy disk drive of a computer and used for storing data.

floppy disk drive The drives in the computer used when storing or retrieving data on a floppy disk.

floptical disk An optical disk that can be erased and reused.

flowchart A diagram showing a logical progression. Flowcharts are used when planning the design of a computer program and serve as a guide as it is developed.

flowcharter A computer program used in creating flowcharts.

flush In word processing, print in alignment along a margin. For example, when aligned with a left margin, lines of type can be referred to as flush left.

font A particular style and design of characters, identified with a name such as Helvetica or Courier.

font card An expansion board, containing additional fonts, installed inside a printer.

font cartridge An alternative to the expansion board, with fonts stored in a cartridge device inserted into an external slot on a printer.

font editor A program that allows the user to modify existing fonts and save them for later use.

font family Multiple variations of a single typeface, such as an italicized version and a nonitalicized version of the Courier typeface.

font number A number in an application program, such as a desktop publishing program, for identifying a specific font.

font size The size of a font, measured in points.

footer Words printed at the bottom of a page. This may include the page number as well as other identifying information such as the author's name.

footprint The amount of desk or floor space required by a piece of equipment, such as a printer or CPU.

forecast Using data-based techniques to make assumptions about the future.

foreground The characters and graphics on a computer display screen.

form   A document, or a screen image, with spaces in which the user is directed to enter specific information.

format   The arrangement of data. For example, word processing programs offer easy-to-use commands for designating the format of text data, with page numbers, double-spacing, and a title page. An alternate use of format relates to the preparation of a disk, with sectors and internal labels, so that it can be used for storing data.

formatting   Indicating the format in which a document is to be printed, including margin size, double or single spacing, and font choice. Generally, easy-to-use formatting capabilities are built into word processing programs.

form feed   Moving continuous feed paper forward in the printer so that the paper is lined up to begin a new page.

form letter program   A software program that enables the user to produce, and address, form letters—a single letter personally addressed to an array of individuals. This technique is also referred to as MAIL-MERGE.

forms program   A software program that enables the user to create forms, such as an order form, on the computer.

FORTRAN The acronym for FORMULA TRANSLATOR, one of the earliest computer programming languages, developed for use in the engineering and scientific community.

486 The abbreviation of the INTEL 80486 MICRO-PROCESSOR, which is currently one of the fastest of Intel's microprocessor family.

4GL The abbreviation of FOURTH-GENERATION LAN-GUAGE, a programming language that is much easier to use than standard, high-level languages like COBOL. Fourth-generation languages employ the terms that humans use in everyday speech, and are generally used with database management systems as a means of retrieving and storing data.

fourth-generation language (See 4GL.)

FPD The abbreviation of FULL-PAGE DISPLAY, a monitor with a display screen shaped like an 8½-by 11-inch page. An FPD is particularly useful when designing a page layout with a desktop publishing program.

FPU (See FLOATING-POINT UNIT.)

fragmentation The presence of small pieces of un-used storage on a storage medium, occurring as a result of continually adding and changing files. If fragmentation becomes widespread, the process of accessing files will be slowed down.

frame A boundary surrounding a graphic image, generally rectangular in shape unless changed by the user, to correspond to the shape of a display screen.

frequency In measuring electricity, the number of cycles an electrical current makes per second, with one hertz representing one cycle per second.

friction feed A method of feeding paper through a printer in which the paper is grasped between two rubber rollers and propelled forward. Friction feed is often used with dot-matrix printers.

friendly Used to describe computer programs and equipment that are relatively easy for users to understand and use. Similar in meaning to USER-FRIENDLY.

front end Another term for USER INTERFACE; a software program that assists users with applications. For example, a user interface might provide a menu from which commands can be chosen or, more likely, icons that the user can select with a mouse.

front panel The area on a computer, printer, or other device in which the various control switches and buttons are located. The front panel is also referred to as the CONTROL PANEL.

fry To destroy electrical equipment by running an excessive amount of current through it.

FTAM The abbreviation of FILE-TRANSFER ACCESS AND MANAGEMENT.

FTM The abbreviation of FLAT TECHNOLOGY MONITOR.

Full Duplex Complete, simultaneous transmission of data over a communications link, such as a telephone line.

full-page display A special monitor used in word processing and desktop publishing applications, which is designed to display a standard 8½- by 11-inch page. This arrangement simplifies the process of developing a page without having to scroll back and forth. Abbreviated FPD.

full-screen A situation in which the complete monitor screen is available for use, rather than a section of it.

full-text searching Searching a complete document to locate specific information, such as the occurrence of certain words or dates.

function A job, or a task, performed by the computer. For example, word processing programs include a mail-merge function.

functional specification The actual, real-world tasks that a computer system will be designed to perform. These tasks must be carefully outlined before the system can actually be developed.

function keys Keys that are used within a program to perform specific functions. Function keys are labeled with an F, followed by a number.

fuzzy logic A branch of logic in expert system and artificial intelligence applications that allows for degrees of uncertainty. With fuzzy logic, an operation does not have to yield an outcome only of *true* or *false,* but also *probably true, probably false,* and variations in between.

G

G The symbol for GIGA or GIGABYTE.

game cartridge   A game stored in read-only memory on a computer chip encased in a plastic cartridgelike case. To play the game, the cartridge is inserted into a computer or other device with a special slot.

Game Control Adapter   A port that can be installed on IBM and compatible personal computers for use in attaching a joystick for playing computer games.

garbage   A slang term for meaningless data.

gas-discharge display   Another term for GAS-PLASMA DISPLAY.

gas-plasma display A technology used in designing the screens of laptop computers, based on the use of neon gas with a flat panel.

gateway A connection that allows a local area network to pass information to another local area network, or to a mainframe computer.

GB The abbreviation of GIGABYTE.

generalized routine A routine in a program that performs a wide range of tasks.

general-purpose computer A computer for a wide range of applications, depending on what software programs are installed on it.

general-purpose language A computer programming language used in developing a wide range of applications. Examples of general-purpose languages include COBOL and BASIC.

gibberish A slang term for unusable data.

giga The prefix that represents one billion.

gigabyte Approximately one billion bytes, abbreviated GB.

gigahertz Occurring approximately one billion times per second, abbreviated GHZ.

glitch A slang term for a problem of some kind in a computer program.

global operation Any function performed throughout a file or document. See GLOBAL SEARCH AND REPLACE.

global search and replace A word processing function that changes a word or phrase throughout a document. The program searches for every occurrence of the word or phrase and makes a substitution in each case.

GOSIP The abbreviation of GOVERNMENT OPEN SYSTEMS INTERCONNECT PROFILE, a version of the OPEN SYSTEMS INTERCONNECTION MODEL (OSI) adopted for use in the federal government. OSI is a set of international standards developed as a basis for connectivity between computers and related devices.

gppm The abbreviation of GRAPHICS PAGES PER MINUTE, the average number of pages containing graphic images that a laser printer can print in one minute. Printing graphics pages is a slower process than printing standard text pages.

graphical user interface A program that uses the computer's graphics capabilities to make applications programs easier. Graphical user interfaces generally represent application program functions with icons the user clicks with a mouse. In the traditional approach, the user must memorize a series of commands to use an application.

graphics A very general term for the use of a computer to create images ranging from very simple line drawings to complex, full-color pictures. These images and pictures may be stored as graphics files and/or printed.

graphics-based A general term to describe hardware or software for creating images and pictures, rather than characters alone.

graphics coprocessor A microprocessor installed on a personal computer to render it more efficient in handling graphics applications. The graphics coprocessor can handle graphics tasks, such as creating images, while the main microprocessor is used for other tasks.

graphics-display A monitor designed to display graphic images generated from a computer. Some monitors can only handle characters and thus display only text and numbers, while others are able to display shapes and images.

graphics file formats Standard file formats used to create and store graphic images. These formats simplify the process of transferring graphics files from one program to another.

graphics for business Graphic images that are especially useful in business presentations, including pie charts and bar graphs.

graphics mode The display mode on an IBM or compatible personal computer in which, not only standard characters, but also graphic images can be displayed.

graphics pages per minute (See GPPM.)

graphics printer A printer designed to print graphic images rather than being limited to characters.

graphics resolution Screen resolution in which the dots are close enough to represent detailed graphic images.

graphics screen A screen designed to operate in graphics mode, displaying, not only standard characters, but also shapes and pictures.

graphics tablet A device, consisting of a flat tablet and stylus, to convert graphic images into input that can be processed as data by a computer. Another term for a graphics tablet is a DIGITIZER.

gray scaling The process of representing an image in an almost unlimited range of gray shades, as in a black-and-white photograph. Gray scaling is used on a monitor without full-color capabilities.

greeking In desktop publishing, the process of previewing a page on the screen, before actually printing it out, with the preview function available in most word processing and desktop publishing programs.

grid Even-spaced rows and columns forming the boxes that serve as the basis for a spreadsheet. These boxes are displayed on the screen of the monitor and contain data meaningful to the specific purpose of the spreadsheet.

gridsheet Another term for SPREADSHEET.

grounding Using electric wiring techniques to prevent the possibility of electric shock to people or damage to equipment.

grouping Assigning elements of data in groups based on characteristics that the elements have in common.

Group 3 protocol The CCITT protocol used for sending facsimile messages.

Group 4 protocol The CCITT protocol used for sending facsimile messages over ISDN-based networks.

groupware Software that helps workgroups to communicate and work together. Features of groupware include electronic mail and meeting schedulers.

guest computer In a network, a guest computer operates under the direction of another computer, referred to as the HOST.

GUI The acronym for GRAPHICAL USER INTERFACE, pronounced "gooey."

gutter In a document arranged in multiple columns, the gutter is the space between each two columns.

GW-BASIC A version of BASIC often provided with DOS.

H

hack   To change a portion of the code in a computer program in clumsy manner, often without adequate understanding of how the program is designed.

hacker   A slang term for someone who is not technically trained to use computers, yet has a serious interest. This person learns by experimentation, which often involves attempting to enter into databases and computer systems without formal authorization. A hacker may, or may not, be interested in obtaining illegal information.

half-card   An expansion card (printed circuit board) half as long as a standard card and designed to fit into a personal computer with limited space in the cabinet. For example, while a slot for the expansion board may be readily available, a hard disk or other internal device may be taking up a large portion of the space above the slot.

half duplex transmission Data communications that occur in both directions, but in only one at a time. An example of half duplex communication is a walkie-talkie, in which only one of the two parties can transmit at a time.

half-height Disk drives in IBM and compatible personal computers can be either full height or half-height. Half-height drives require less space.

halftone On a monitor, halftone images are displayed in shades of gray, based on black and white dots.

halt instruction A command that immediately stops the execution of a command.

hammer The mechanism on an impact printer that actually strikes the ribbon to cause the imprint of a character to appear on the paper.

hand calculator A small calculator, used for basic arithmetic calculations, that can be held in the palm of the hand.

hand-held computer A small computer that can be used to perform basic tasks, including calculations and data storage in a simple database.

handle A graphics term referring to an outline of a graphic image that can be selected with a mouse for subsequent use.

handshaking The ability of two different computers, based on different designs, to communicate with each other. Handshaking is accomplished through computer programs containing special protocols.

hands-on A slang term for experience derived from actual use of a computer system, rather than from reading about it in a textbook.

handwriting recognition The ability of the computer to recognize a user's handwriting, either as a means of inputting data or as a security precaution. Handwriting recognition requires the use of a special scanning device.

hang A system problem that results in the computer not responding to commands issued from the keyboard. Generally, the computer will respond only after it has been rebooted.

hanging indent Used to describe the first line of a hanging paragraph, in which all lines but the first are indented.

hanging paragraph A paragraph in which all lines but the first are indented.

hard Anything that physically exists, like data printed on paper, as opposed to being soft, i.e., conceptual.

hard card A hard disk installed as an expansion board, similar to a graphic interface card or communications board. Hard cards do not have the storage capacity of a standard hard drive, but they are much easier to install.

hard coded Instructions programmed into hardware, or software that cannot be modified in any way.

hard copy Data or graphic images printed on paper and thus in physical existence. For example, a document that has been printed on paper is hard copy, whereas one that is only displayed on the screen is soft copy.

hard disk A magnetic disk that stores data and is permanently mounted inside the computer. Hard disks hold much more data than floppy disks, and the data are not subject to damage from careless handling as with floppy disks.

hard disk drive The mechanism that writes on, and retrieves data from, a hard disk.

hard error An error, resulting from a failure in a hardware device, that may or may not bring processing to a complete halt.

hard failure A system failure that results from a hardware, rather than a software, problem. Generally, recovery from a hard failure requires that the hardware be repaired or replaced.

hard hyphen A hyphen required by a rule of grammar.

hard return In word processing, an action to force the beginning of a new line. For example, at the end of a paragraph, a hard return begins a new paragraph at the next line. Or, when creating a list, a hard return indicates the beginning of the next item on the list, on the next line. Generally, a hard return is created by pressing the Enter key.

hardware The CPU, peripheral devices such as printers and modems, cables, cords, and other equipment used for information storage and management.

hardware-dependent Programs or computer languages that are only useful on particular types, or brands, of hardware. For example, the DOS operating system is hardware-dependent in that it is designed for use on IBM and compatible personal computers, as opposed to Apple computers.

hardware failure Another term for HARD FAILURE.

hardware key A metal key that fits into a lock on the front of many personal computers to completely disable the system and prevent access by unauthorized users.

hardware resources Actual hardware as well as what these resources yield to the user, including storage space for data, CPU time, and processing power.

hardwired A computer or other hardware device that is connected to a network through a cable or similar physical connection.

Hayes compatibility The ability of a non-Hayes modem to understand commands issued to a modem that has been manufactured by Hayes Microcomputer Products. Hayes products are used so extensively that they are considered a standard that other manufacturers often emulate.

HDBMS The abbreviation of HIERARCHICAL DATABASE MANAGEMENT SYSTEM, a now-outdated method of database management for mainframe computers in which the data are organized in a hierarchical structure. This method has been supplanted by relational and object-oriented models.

HDTV The abbreviation of HIGH-DEFINITION TELEVISION, a new technology currently being developed that will yield much higher quality video images than the current technology.

head The device that writes data on, and retrieves them from, a disk drive.

head-cleaning device A tool that applies a cleaning fluid to a magnetic storage device to remove dust.

head crash A slang term for the failure of a head to operate, often caused by dust. The result of a head crash may be a scratch or other damage to the disk, and unrecoverable loss of data.

header Text that appears at the top of each page of a document. The header often includes the author's name or identifying information about the document, as well as a page number.

header record The first record in a group of records from a database.

helical-scan cartridge A storage medium involving the use of magnetic tapes installed in cartridges, similar to the tapes used with a VCR.

help Assistance provided to the user, through information the user can request during the course of using a software program, or by reading associated documentation.

Help key A key that can be pressed to receive additional information while using an application. The Help key is labeled as such on the Apple Macintosh keyboard, while, on the IBM, one of the function keys (usually F1) provides this feature. Not all application programs are designed to actually work with the Help key and may indeed ignore it.

help screen A screen in an application program that provides the user with additional information on one of the features in the program. Often, a help screen is displayed after the user has requested this information by pressing the HELP KEY.

Hercules Graphics A system designed for displaying graphic images on IBM and compatible personal computers. Hercules Graphics are added to a system by installing a Hercules Graphics card. Use of this system is so widespread that it is often emulated by other graphics systems companies.

hertz Number of cycles per second.

heterogenous network A computer system composed of units from various vendors that are most likely incompatible with each other yet are able to exchange information across a network that connects them all together. Many large organizations have heterogenous networks connecting computers from manufacturers such as IBM, Digital Equipment Corporation, and Hewlett-Packard.

Hewlett-Packard A major manufacturer of computer hardware and peripheral equipment, including midrange computers, high-power workstations, and laser printers.

hex The shortened version of HEXADECIMAL.

hexadecimal Any number from the hexadecimal numbering system, which is based on sixteen symbols, the numbers 1–9 and the letters A–F. The hexadecimal numbering system is used as an alternative to the binary system in computer applications.

HGC The abbreviation for Hercules Graphics card. (See HERCULES GRAPHICS.)

hidden file A file not normally readable or modifiable by users. Certain operating system and application program files are hidden so that users do not accidentally make changes that would cause damage to the program or to data.

hierarchical Based on a pyramid design, like a family tree, with layers of interconnected elements. The hierarchical model has been used extensively in the computer industry, particularly in the area of database design.

hierarchical database management system (See HDBMS.)

high-definition television (See HDTV.)

high-density disk A floppy disk that holds more data than either a single- or double-density disk. A 5¼-inch high-density disk holds 1.2 megabytes of data, and a 2⅓-inch high-density disk holds 1.44 megabytes of data.

high-density television   (See HDTV.)

high-level language   Computer programming languages that use commands and terminology similar to human language. Languages such as COBOL and FORTRAN are considered high-level. This can be contrasted with low-level languages, which are written in binary code, for example, and are thus much closer to direct interaction with the machine. High-level languages must be processed through a compiler before they can actually be put to use.

highlight   To make data on the monitor screen easier to see by making it brighter or making it blink. For example, the cursor often blinks to aid the user in locating it.

high memory   Generally, the area of memory occupied by the operating system and thus unavailable to the user for storing data.

high resolution   The quality of image produced on a monitor screen, or by a printer. Images on the screen are made up of tiny dots called pixels. The closer together the pixels are, the sharper and clearer the resulting image. Printers also use dots, and, by the same principle, the more dots per inch, the higher the resolution that results.

high storage   Another term for HIGH MEMORY.

high tech   Innovative developments in computer science and engineering.

homebrew   Computers and related hardware that have been built either at home or within an organization, and not purchased from a manufacturer. HOME-GROWN is a similar concept applied to SOFTWARE.

home computer   A personal computer, or microcomputer, used in the home.

home-grown software   Software developed by in-house staff as opposed to being purchased from a computer store or software vendor.

Home key   A key that automatically takes the cursor to the home position on the screen, which is generally the upper left-hand corner.

home management software   Software programs that perform functions relevant to the home, such as budgeting and checkbook programs.

host In a large mainframe computer scenario, the host is the main computer in a network, usually containing the organization's major applications and most critical data. The host can also refer to any computer that is accessed by users at other sites.

hot key A key that, within a program, is assigned by the user to perform a specific function. For example, some software programs and operating systems, including DOS, allow the user to create a small program and then assign a specific key to execute it each time it is pressed. This can save time, because the user does not have to enter a series of commands.

hot link A direct link established between two applications, such as a database program and a word processing program, so that, when changes are made to a copy of a file in one program, the same changes are automatically made to the same file in the other program.

hot site A computer system established in another location with the same hardware and applications. Large organizations often establish hot sites so that, in the case of a disaster that brings down the system, processing can continue at the alternate site.

housekeeping Computer-related tasks that serve to keep the system maintained, such as deleting unnecessary files from the hard disk.

HP The abbreviation of HEWLETT-PACKARD.

HP-compatible printer A laser printer that can respond to commands in Hewlett-Packard's printer control language (PCL), issued by applications such as word processing and desktop publishing programs to control Hewlett-Packard printers. Because Hewlett-Packard printers are in such widespread use, other printer manufacturers have built this compatibility into their printers.

HPGL The abbreviation of HEWLETT-PACKARD GRAPHICS LANGUAGE, a graphics file format developed by Hewlett-Packard.

human engineering The science of designing hardware and software products that work with, and enhance, the natural capabilities and limitations of human beings.

human factors engineering Another term for HUMAN ENGINEERING.

human language The language that human beings speak, and write, as opposed to computer language.

human–machine interface   The point at which the human being and the machine come together, as in the icon-based graphic interface being used with many computer programs.

HyperCard software   Software developed for the Apple Macintosh computer that allows the user to develop HYPERTEXT applications. HyperCard software basically treats data like cards with, for example, a graphic image on one card and sound on another. The cards are arranged in stacks. These cards can then be retrieved, and combined, in interesting and creative ways within the application.

hypermedia   A broad application of HYPERTEXT, which focuses on the integration of various media such as sound, graphics, and video. Hypermedia applications are designed to parallel the random associations that occur in human thought.

hypertext   An innovative database system in which various multimedia elements, including text, graphics, video, and sound can all be linked together. A wide array of multimedia applications, particularly in areas such as education, are becoming available. Hypertext is available both on Apple and IBM personal computers.

hyphenation   In word processing, hyphenation refers to splitting a word that is positioned at the right margin according to its syllables, so that part of it is moved to the next line.

Hz   The abbreviation of MEGAHERTZ.

I

I-beam pointer A cursor/pointer used for text input when the primary input device is a mouse, particularly in word processing and database applications. For example, the standard arrow that appears on the screen when a mouse is being used will appear as an I-beam in the areas where text is required.

IBM Corporation A major manufacturer of hardware and software products, from large mainframe computers to personal computers, and software that runs on these computers.

IBM Personal Computer A general term for the family of personal computers manufactured by IBM, including the PC, AT, and XT, as well as the more powerful PS/2 line of computers. IBM Personal Computers use microprocessors from the Intel Corporation. Software designed to work with the IBM line of personal computers will generally also run on IBM-compatible computers.

IBM Personal System/1 IBM's much more advanced and powerful line of personal computers, based on an Intel 80286 microchip and personal computer. This line generally makes use of IBM's OS/2 operating system rather than DOS. The standard abbreviation is PS/2.

IC The abbreviation of INTEGRATED CIRCUIT, a complete electronic circuit installed on a silicon chip.

ICCA The abbreviation of INDEPENDENT COMPUTER CONSULTANTS ASSOCIATION, a professional association of individuals who provide computer consulting services.

ICCE The abbreviation of INTERNATIONAL COUNCIL FOR COMPUTERS IN EDUCATION, an organization of individuals involved in providing computer instruction at the elementary and secondary levels.

icon A small image on a display screen that, in conjunction with a mouse, selects and executes program functions. Icons form the basis of a graphical user interface.

iconic interface A graphical user interface (GUI) based on the use of icons that allow the user to select functions with the use of a mouse rather than typing commands. (See GUI).

IDE interface The abbreviation of INTELLIGENT DRIVE INTERFACE, a hard disk with an integrated controller used in IBM and compatible personal computers.

identifier A name, or symbol, that applies to a variable in a program, or a program itself.

idle time Periods of time during which the computer system is available but not in use.

IEEE The abbreviation of INSTITUTE OF ELECTRICAL AND ELECTRONICS ENGINEERS, an association of engineers and scientists whose activities include setting standards for computers and data transmission.

IFIP The abbreviation of INTERNATIONAL FEDERATION OF INFORMATION PROCESSING, an international association of organizations representing information-processing professionals.

IGES A graphics file format for three-dimensional images.

illegal character A character not recognized by the program.

image A shape or picture developed through a graphics software program and meant to represent a specific object, such as a building or a person, or a concept, such as a statistical trend. Also, an exact duplicate, of an insurance form or an x-ray, for example, that is stored in the computer's memory.

image processing The process of creating, storing, and manipulating images, particularly exact duplicates of images such as legal forms, photographs, or x-rays.

ImageWriter A family of dot-matrix printers developed by Apple Computers and used with the Macintosh.

imaging Another term for IMAGE PROCESSING.

immediate access The ability to quickly retrieve information from a computer; also another term for RANDOM ACCESS.

impact printer Printers that operate by striking a hammer with embossed type against a ribbon that, in turn, strikes the paper. Dot-matrix and daisy wheel printers fit into this category.

import To bring a file into one program that was originated in another program. For example, creating a document, and then importing it from a database into a word processing program.

inactive A command that has been entered and is waiting to be processed.

inclusive OR operator In Boolean logic, an operation that yields a value of true if one of the statements on either side of the OR is true.

increment To add a specific amount to the current value. For example, to add 10 to the current value is to increment it by 10.

incremental backup A process in which only the most recently used files since the last backup are backed up, instead of a complete disk. For example, if a disk containing word processing backups has previously been backed up, and additional changes are made on only one file, then only the file that was changed since the last backup would be copied.

indent   To begin a line, or a paragraph, one or more spaces away from the left or right margin.

indentation   The blank space that results from the use of the indent function.

Independent Computer Consultants Association   (See ICCA.)

independent software vendor   A company that develops and sells computer software programs.

index   A means of identifying a piece of data that has been arranged in a database. An index may be a number, letter, or symbol, and is used by the program to identify data during the process of making retrievals. Also, the process of organizing the information contained in a document, like the index in the back pages of a book.

indexer   A software program that creates indexes for large documents.

indicator   A device, such as a small light, that shows the current status of a hardware device. For example, printers are equipped with lights and panels that indicate that the machine has been turned on and a font selected.

industrial data collection device   A device, in industrial situations, that measures the time employees actually spend in their jobs.

industrial robot   A mechanical device that can be programmed to perform manufacturing-related tasks such as assembling or moving parts.

Industry Standard Architecture (ISA) Bus   The underlying design for making all of the parts of a personal computer work together, used in older models of the IBM PC, including the PC/AT and PC/XT.

inference   Drawing a conclusion based on a set of data that are assumed to be valid.

inference program   A software program that collects and analyzes data and then makes inferences based on parameters provided by the user.

information   Though often used interchangeably with the word *data,* information is actually meaningful and useful to human beings, while data are merely raw facts stored in a computer.

information center   A centrally located area in a large organization where users can come to receive training in computers as well as, to a lesser extent, to access information based on CD-

ROM discs and through online information services. Information centers were popular during the early to mid-1980s. With the increasing availability of personal computers and growing computer knowledge among users, many information centers are being disbanded.

information management   Broadly used, the process of organizing and storing an organization's valuable information resources and making sure users have needed access.

information network   A network of computers based at libraries spread out among geographical areas and connected through telecommunications facilities to share information.

information processing   The use of computers for storing and modifying information, and for making it available to users.

information processing center   A computer center, where computers are used to store and process information.

information processing curriculum   Another term for COMPUTER SCIENCE CURRICULUM, where students learn the basic fundamentals of how computers work, as well as to become proficient in programming languages.

information resource management Another term for INFORMATION MANAGEMENT.

information retrieval The process of searching for information that has been stored in a computer.

information science The study of how information is compiled and disseminated.

Information Services The name many companies use to refer to their computer department, rather than Data Processing or Management Information Systems. Information Services implies that actual services are being provided to users, which is the mandate that the computer staff has received from its executive management. The abbreviation of Information Services is IS.

information system The hardware equipment, software programs, and manpower associated with an organization's computer-based information strategy.

information technology Broadly, technological developments that provide better computers and telecommunications services.

inhibit To stop something from happening.

in-house training Training that occurs in the organization, often under the guidance of its own training staff.

init A small program in a Macintosh that is executed when the computer is turned on.

Initial Graphics Exchange Specification (See IGES.)

initialize To set a variable in a computer program at a certain value so that the value is the same each time the program begins execution.

initial program load The action taken by the computer, after it has been turned on, to load the operating system into main memory. The initial program load is part of the START-UP process.

ink cartridge A disposable container of ink used with an INK-JET PRINTER.

ink-jet printer A printer based on ink-jet technology, in which ionized ink is sprayed onto the paper. Ink-jet printing produces high-quality documents, similar to laser printing though generally at a lower cost. Ink-jet printers that print in full color are available.

input The process of entering data into a computer from a keyboard or other device. The data actually entered are also referred to as input.

input data Another term for the noun form of input, which is data that have been entered into the computer and are ready for processing.

input device The piece of equipment used to enter data into a computer, including the keyboard or graphics tablet.

input/output A general term referring to the process of entering data into a computer, or retrieving them. Input/output can refer to devices, to programs, and to processes. Input/output is abbreviated I/O.

input–output statement A line of instruction in a computer program that guides the movement of data from, for example, main memory to the display screen or to a printer.

inputting The process of entering data into a computer.

inquiry A request for information in an application program or from a database management system.

insert To place a character between two existing characters, as in word processing.

insertion point   The position at which an insert occurs.

insert mode   A mode in word processing in which any character entered will automatically be inserted between characters to the right and left of the cursor.

Ins key   The key that activates and deactivates the insert mode in a program.

installation program   A small program designed to install another program. Most software products include such a program, which guides the user through the process of loading the product in main memory for the first time, designating the printer and monitor the software will operate with, and naming the directory on the hard disk where the software will be stored.

Institute of Electrical and Electronics Engineers   (See IEEE.)

instruction   A command the user gives the computer to initiate a task.

instructional computing   A generic term for the use of computers in the teaching process.

integer   Any whole number or zero.

integrate To use hardware devices and telecommunications techniques to connect diverse computers, such as those manufactured by IBM and Apple, together so that they are able to operate as if as one.

integrated circuit (See IC.)

integrated drive electronics (See IDE.)

Integrated Services Digital Network A set of international standards that guide the establishment of networks in which different types of data, including voice, information, and video, are sent over telephone wires.

integrated software Software that consists of various component applications, such as word processing and spreadsheet, that work together and are sold as one product.

integrity In reference to data, the purity, or cleanliness of the data available for use in the applications for which it was created.

Intellifont A name given to font technology developed by Hewlett-Packard for use with their Printer Control Language.

intelligence   Used in describing a hardware device that is capable of processing information (see also INTELLIGENT TERMINAL).

intelligent terminal   A monitor and keyboard with processing capabilities built in, rather than simply connected to a computer, such as a mainframe. Examples of intelligent computers include point-of-sale terminals that not only act as cash registers but also capture and process data, which is in turn fed to a central computer.

Intel Microprocessor   Microprocessors manufactured by the Intel Corporation. Intel microprocessors are used in all IBM and compatible personal computers.

intensity   The level of brightness of the data being displayed on a monitor. This can be adjusted with a switch or button that controls this condition.

interactive   Able to respond to commands from a person, in two-way communication, with each command followed by a response. Database programs, in which users enter commands and the program responds with information, are interactive. Word processing programs are also interactive in that they respond immediately to commands and data from the user.

interactive graphics The development of graphic images through the use of a software program that allows the user to modify each image over time until satisfaction is attained.

interactive session A period of time during which the user is directly involved in operating the computer and engaging in activities such as retrieving, modifying, and storing data.

interface A connecting point between two different elements, including computers, devices, programs, and human beings. A communications interface could be a connecting device between two computers.

interface board An expansion board for a personal computer that is used for connecting external devices such as printers.

interference Noise and other unwanted distortions, over a communications line, caused by stray signals.

interlacing A technology used in monitors to provide greater screen resolution.

interleaving A means of managing the main memory of a computer by dividing it into sectors that are accessed alternately by the CPU. By alternating this access, the overall speed of the CPU is enhanced.

internal clock A chip-based clock inside a computer that keeps track of the current time.

internal command DOS commands, used for DOS operations such as COPY and stored in the COMMAND.COM file.

internal font A font that is built into the actual printer hardware rather than software driven. Most printers have one or more resident fonts.

internal memory Another term for MAIN MEMORY.

internal modem A modem installed inside a personal computer on an expansion board, rather than separate and connected with a cable.

International Council for Computers in Education (See ICCE.)

International Federation of Information Processing (See IFIP.)

International Standards Organization An international organization responsible for developing standards for telecommunications and data exchange.

Internet   An international network connecting government agencies, universities, and other organizations for the purposes of sharing information.

interpreter   Similar in function to a COMPILER, an interpreter translates program language into machine code and immediately executes it. A compiler translates the program into machine code but does not actually execute the program.

interrupt   A signal activated when an event of some type occurs that, as a result, stops a program from executing. For example, a printer may send an interrupt to a word processing program if an error occurs that prevents the printer from completing a task. The user is then informed of this event with a screen message.

inventory control   A software application in a business setting in which inventory storage is monitored by a computer application.

inverse video   A method of displaying information on a monitor screen in which dark characters appear on a much lighter background. Some users prefer this method. Also referred to as REVERSE VIDEO.

invoke To issue a command or otherwise activate a function in a program.

I/O The abbreviation of INPUT/OUTPUT.

I/O board A circuit board that manages incoming and outgoing data between the computer and devices such as printers.

I/O port A port, or connection, on the computer that is used to establish a path for the data passing between the computer and devices such as the monitor and disk drives.

IPL The abbreviation for INITIAL PROGRAM LOAD.

IRMA board An expansion board for connecting personal computers to mainframe computers. Personal computers with IRMA boards can act as personal computers as well as receive and transmit data to a mainframe computer.

IS The abbreviation of INFORMATION SERVICES.

ISA Bus The abbreviation of INDUSTRY STANDARD ARCHITECTURE BUS.

ISDN See INTEGRATED SERVICES DIGITAL NETWORK.

ISO   The abbreviation of INTERNATIONAL STANDARDS ORGANIZATION.

isolation   Information kept separate from the rest of the system to protect it from unauthorized users. This may be contained in a database restricted from general use, or a completely separate computer may be in operation to accomplish this task.

ISV   The abbreviation of INDEPENDENT SOFTWARE VENDOR.

italic   A font style in which all characters are slanted to the right.

item   A choice on a program menu that may be selected by a user.

iterate   In programming, when an action occurs repeatedly until a process is completed. For example, a banking program might process all debits in the same manner until all debits have been processed.

iteration   Each occurrence of a repetitive process.

iterative   A process in which a repetitive action is performed.

J

jack A device used to connect electrical devices such as cables and wires.

jacket A paper or plastic cover to protect a disk from dust and scratches.

jaggies Jagged lines in a graphic image as it appears on a monitor screen, particularly on curves and in corners. These jagged lines are usually the result of being displayed on a monitor that does not have high enough resolution to display fine lines (see also ALIASING).

jargon The slang-laden language spoken by computer professionals and those who aspire to their ranks.

JCL The abbreviation of JOB CONTROL LANGUAGE, a set of commands used with mainframe computer operating systems to guide the execution of a specific task.

job A task performed by a computer under the direction of a computer program.

job control language (See JCL.)

job number A number assigned by the user, or the computer system, to identify a task being performed by the computer. This number enables the user to monitor the progress and completion of the task.

job queue A line of jobs, or tasks, waiting to be executed by the computer. Generally, the first job in line is completed first.

job scheduler A computer program that monitors and controls the scheduling of jobs; generally used with mainframe computers. For example, a job scheduler might schedule large and less critical jobs to be completed during the night hours, when fewer users are competing for the system.

job statement One line of the series of statements used in job control language.

job turnaround   The time that passes from when a job is submitted to the computer system to when it is actually completed.

join   In relational database management, a retrieval operation that involves data from two or more databases. For example, a retrieval of an employee list that requires job titles from one database and home addresses from another database is a join operation.

join operation   (See JOIN.)

joystick   A leverlike device used like a mouse, particularly with computer games. Unlike a mouse, the joystick device stays stationary. The lever shifts in all directions to move the cursor around the screen.

Julian number   A date represented by the year and the number of days that have passed within the year up to that day. For example, February 10, 1993, in its Julian format, is 93-41, because, on February 10, 41 days have passed. Some computer programs make use of Julian dates.

junk   Another term for data that have somehow been damaged or compromised.

justification Arranging text so that it is in alignment along the right or left margin, or both.

justify The process of aligning text along the right or left margin. In word processing programs, this is generally accomplished with a simple command.

K

K The abbreviation of KILO. Kilo represents 1,000 in a decimal system and 1,024 in a binary system. Kilo is sometimes also used to refer to kilobyte in computer terminology.

kb The abbreviation of KILOBYTE, which is 1,024 bytes.

kc The abbreviation of KILOS PER SECOND, generally associated with data transfer speeds.

Kermit A protocol used to send data with modems over telephone lines, developed at Columbia University. Kermit is not widely used in the home computer setting.

kernel Programs in an operating system that perform the most basic functions.

kerning Adjusting the white space between two characters so that, when printed, they are closer together and appear more in proportion to the other characters.

key A button on a computer keyboard. Key can also be a verb, as in to key in data, instead of "type in" or "enter" data.

keyboard An input device designed similarly to a typewriter and used to enter data into a computer. The alphanumeric keys of the computer keyboard are in the same position as those on the typewriter; however, additional function keys are also available to the user.

keyboarding Like the word *key,* a verb for the process of entering data into a computer.

keyboard templates A piece of plastic or paper that can be slipped over a section of the keyboard, usually over the function keys, and shows commands associated with each specific function key. For example, many word processing programs, when purchased, include a template that the user can tape over the function keys to show the commands for that program.

keyboard terminal A keyboard that is generally connected to a mainframe computer. The user keys in data on the keyboard and sends it to the mainframe.

keyboard-to-disk system A computer system that allows the user to use a keyboard to type data directly onto a storage disk.

keyboard-to-tape system A computer system that allows the user to use a keyboard to type data directly onto a storage tape.

keypad A small keyboard, either positioned on the right side of the standard keyboard or a separate device, containing numerical keys and possibly function keys. The keypad is particularly useful for numeric operations in that it can be used like a calculator.

keystroke Pressing a key on a computer keyboard. Data entry speeds are generally measured by counting the number of keystrokes in a set amount of time.

keyword In database management, certain fields on a record that can be used to identify that record for retrieval. For example, on a personnel record, the last name field, or the social security number, can be used as keywords to locate and retrieve a specific employee's record.

kHz The abbreviation of KILOHERTZ, 1,000 bytes per second, a measurement used for data transmission speeds.

kill To delete data or to stop a command during execution.

kilo (See K.)

kilobyte (See KB.)

kilocycle One thousand cycles.

kilohertz (See KHZ.)

kilomegacycle One billion cycles.

kinematics A CAE (computer-aided engineering) process used in designing machinery, which animates moving parts so that the operation of the machine can be demonstrated.

kludge A computer system difficult to use due to poor planning and design.

knowledge acquisition The ability of computer software to store and use information that will enable it to make humanlike decisions.

knowledge base   A database that contains facts and statistics pertaining to a specific subject.

knowledge engineering   A branch of engineering concerned with designing computer systems to solve problems that generally require a high level of human reasoning.

knowledge industries   Software and computer companies that supply information products to other industries.

knowledge work   Broadly, "white-collar" occupations that primarily require the ability to process information and make decisions.

knowledge worker   An individual who performs knowledge work.

L

label A name used to identify a file, a part of a program, or a storage medium such as tape.

lag A time delay between two events.

LAN The abbreviation of LOCAL-AREA NETWORK, a computer system consisting of personal computers, within a department or office, that are connected to each other with cable and are able to share data and application programs.

landscape In word processing, a page that is printed sideways, generally on long paper, such as a spreadsheet. Landscape printing is also used for newsletters that are printed in columns and then folded.

language   A means of communicating, consisting of a system of words and symbols understood to both parties.

language compiler   Another term for COMPILER, a program that translates program code written in human-language–oriented programs such as COBOL into code that can be understood by the computer.

language translation   The process of translating one language to another by, for example, a language compiler.

language translator program   Another term for LANGUAGE COMPILER.

laptop computer   Describes the broad range of personal computers that can be placed on the user's lap. These computers are small, self-contained units, ranging from approximately 5 to 15 pounds, and can operate on battery power.

large-scale integration   The installation of a large number of integrated circuits on a single silicon chip.

laser printer   A printer, based on laser technology, that produces high-quality print with a single laser beam. Laser printers use a rotating drum rolled through toner, which traverses the areas of paper touched by the laser beam. Rather than printing one line at a time, as do impact printers, laser printers print one page at a time.

LaserWriter A laser printer manufactured by Apple Computer and designed to work with Macintosh computers.

LAWN The abbreviation of LOCAL-AREA WIRELESS NETWORK, a local-area network based on wireless technology, in which data are transferred through radio waves as opposed to going through cable.

layering In graphics applications, linking together drawings that are part of one large drawing. For example, a very complex design can be developed and viewed in sections, so that the pieces can be developed in tandem.

layout Arranging text and graphics in a logical and eye-pleasing arrangement, such as in a newsletter or brochure. Desktop publishing programs provide features that allow the user to design and view layouts on the screen of the monitor.

LCD The abbreviation of LIQUID CRYSTAL DISPLAY, a screen technology based on light reflected against a liquid crystal substance and often used in small portable computers and calculators.

leader An area of blank tape at the beginning of a storage tape.

leading edge The edge of the paper that passes through first when a sheet of paper is being fed through a scanner or through a fax.

leaf The last or bottom element in a tree or hierarchical structure.

learn mode The mode in which a computer program is being modified with additional user-written programs, called MACROS, and is thus "learning" to meet the user's specific needs.

lease It is a common practice in the computer industry, particularly as regards mainframe computers, to lease, rather than purchase, hardware and software products. This protects a company's investment if its needs change, or if the technology is outdated.

LED A small electronic display that lights up when a volt of electricity passes through it; usually used for control panels on equipment such as computers and printers.

left justify To vertically align each line of text with the left margin of a page.

legend An explanation of the numbers and symbols used in a chart.

LEN A relatively simplified means of designing large IBM-based networks, without mainframes, in which all machines can communicate with each other.

length The number of characters in, for example, a field on a database record, a label or name, or a document.

letter quality Typewriter quality, or high enough in quality to be compared to output from a standard typewriter.

letter quality printer A printer capable of high-quality output. Generally, laser and ink-jet printers are considered letter quality, while even the best quality print from a dot-matrix printer is considered near letter quality.

LF key The key on a printer that will advance the paper one line at a time.

library A collection of smaller programs, or routines, associated with a large application program. A library can also be a collection of programs available for use by the users of a computer system.

library automation The use of computer technology within a library setting, as a means of managing the card catalog, for example, and providing computer-based information for library users.

library manager A computer program that manages the storage and use of a library of programs.

light bar When menus are displayed on a computer screen, the user often makes a choice by moving a highlighted area, the light bar, up and down the list with the arrow keys.

light-emitting diode (See LED.)

lightness In applications that include color, the amount of light in a specific color.

light pen A light-sensitive electronic device, shaped like a pen, that can be used like a mouse to draw on, and make selections from, a computer screen.

LIM memory A technique used for adding memory to personal computers with the DOS operating system.

line drawing A basic drawing that consists of an outline of a shape, without detail or color.

line editor A basic type of editing program that allows the user to make changes to one line of text at a time.

line feed (See LF KEY.)

line number Many programming languages, including BASIC, assign an identifying number to each line of code.

line of code A statement in a computer program, generally requiring one line of text and assigned a line number.

line printer A fast, yet relatively low print quality, printer. Often, line printers are attached to mid-range or mainframe computers and used for high-volume printing such as invoices or checks.

line printer terminal In computer operating system commands, the printers attached to the system. This is a general term and, in operating system terminology, can refer to any type of printer.

line speed   The speed at which data can be sent over a communications line.

lines per minute   The time a printer takes to print one line of code, and a way of measuring a printer's speed. The abbreviation is LPM.

line surge   A quick and unexpected increase in the amount of voltage currently passing through an electronic device. Line surges often occur at home, where an appliance such as a refrigerator suddenly causes a spike of electricity that affects other appliances and can cause permanent damage to a personal computer.

link   A connection between two devices that allows data to be passed between them. Also, the ability of one spreadsheet to pull data from another one.

link edit   To use a linker program to link various programs together.

linker   A computer program to link other programs together. With a linker, pieces of a program, called OBJECTS, can be developed separately and then combined into a useful application program.

liquid crystal display   (See LCD.)

liquid crystal shutter printer A printer that uses liquid crystal technology and produces print quality comparable to that of a laser printer.

LISP The acronym for LIST PROCESSOR, a programming language that works well with processing lists of data and is often used for artificial intelligence applications.

LISP machine A computer designed especially to run programs written in LISP.

list To organize data in a logical order, such as numerically; also, a verb meaning the process of printing the contents of a database.

listing A printout of a program or data. A listing is generally assumed to be an as-is printout, without any special order or formatting.

list processing Techniques associated with the processing of lists of data.

list processing languages Languages such as LISP that were developed for processing lists.

live data Real data, as opposed to test data, being submitted for processing by the computer.

load To install a software program, or to read a program that already exists in storage into main memory so that it can be used.

loader The function in an operating system that brings programs from storage into main memory.

load sharing The ability of two or more computers to work together to handle the excess processing requirements that often occur during the busiest times of the work day.

local The programs and data files on one's own computer, as opposed to those installed elsewhere and thus remote.

local-area network (See LAN.)

local-area wireless network (See LAWN.)

local network Another term for a LOCAL-AREA NETWORK. (See LAN.)

LocalTalk The cable strategy for connecting Macintosh computers together as a local-area network.

location An area of storage in a computer.

lock   To make a file secure from modification by two users at the same time. (See FILE LOCK.)

log   A record of events associated with a device or program. For example, a system log would include information about who used the system, what they did, and when these events occurred. Logs are often used in tracking problems or unauthorized access.

logging off   The process of ending a computer session.

logging on   The process of beginning a computer session.

logical   The way that users view an application or database design, rather than how it is actually designed from a technical viewpoint. For example, a database containing personnel records must make logical sense to the users while also, from a technical standpoint, be designed to work within the constraints of storage and data processing specifications.

logical operator   Another term for BOOLEAN OPERATOR, used in Boolean logic.

Logical Unit Type 6.2   A strategy developed by IBM for use by large organizations to connect computers so that they can share data.

log in In a large computer system, the process that a user goes through to begin using the system. Generally, this involves entering some type of identification, followed by a password. Using personal computers is much simpler and usually requires merely flipping the machine on.

log off To stop using a computer, generally by exiting from programs and, on a large system, using a command to indicate to the system that the usage is ending. When ending a session on a personal computer, this generally requires little more than exiting the program and turning the computer off.

log on Another term for LOG IN.

log out Another term for LOG OFF.

look-alike A concept similar to LOOK-AND-FEEL, with a program developed to imitate the operation of another.

look-and-feel A similarity between one vendor's software product and another's. While this can be comforting to the user, it can also result in lawsuits between the two companies for copyright infringement.

loop   A section of a program repeated over and over again, with each repeat referred to as an ITERATION.

Lotus 1-2-3   A popular spreadsheet program developed by the Lotus Corporation.

Low Entry Networking   (See LEN.)

lowercase   Noncapitalized, or "small," letters.

low-level format   The first of two formatting processes that a hard disk undergoes, generally at the manufacturer's, before it is ready for data and program storage.

low-level language   A programming language, such as ASSEMBLER, that communicates directly with the hardware. High-level languages, such as COBOL, are closer to the way in which human beings communicate.

low-res graphics   A slang term for LOW RESOLUTION.

low resolution   Lower-quality screen resolution, resulting from a low number of the dots, or pixels, used in creating the images. With low resolution, the images will not be sharp but have jagged edges.

LP The abbreviation of LINE PRINTER.

LPM The abbreviation of LINES PER MINUTE.

LPT The abbreviation of LINE PRINTER TERMINAL.

LQ The abbreviation of LETTER QUALITY.

LSI The abbreviation of LARGE-SCALE INTEGRATION.

luggable A portable computer that can be moved from place to place but is heavier than a notebook or laptop computer. Generally, luggable computers weigh approximately 17 pounds.

luminance The relative brightness of the screen of a computer monitor.

luminance decay The gradual decline of brightness, over time, on a computer monitor.

lunchbox A style of design for a laptop computer, with a cover that folds up and down like the cover of a lunchbox, with the screen inside.

LU6.2 The abbreviation of LOGICAL UNIT TYPE 6.2.

M

M The abbreviation of MEGA, which means one million.

Mac The slang term for Apple's MACINTOSH COMPUTER.

machine address Another term for ABSOLUTE ADDRESS.

machine code Programming code that is directly recognizable by the computer, and does not require compiling. Machine code is also referred to as "low-level" code. Machine code is another term for MACHINE LANGUAGE.

machine dependent Software that will run on only one brand of computer, for example, a word processing program that will only run on IBM or compatible machines.

machine error An error caused by the computer, as opposed to the software.

machine independent Software that will run regardless of the machine on which it is installed.

machine instruction A line of code that can be understood and executed by the computer.

machine language Another term for MACHINE CODE.

machine readable Any data that can be recognized by the computer. Machine readable data include data stored on a hard or floppy disk, a tape, a CD-ROM, or on any number of other media.

machine sensible Another term for MACHINE READABLE.

Macintosh Computer A personal computer designed and manufactured by Apple Computer, known for its easy-to-use, icon-based user interface, graphical capabilities, and wide range of applications for business, home, and education.

macro   A "shorthand" means of writing a series of commands to perform an often-used function. The series of commands can be written once and then, using an application program's macro feature, the series can be given a macro name that, when entered by the user, will cause the much longer series of commands to be executed. For example, a word processing program user might develop a series of commands for merging a letter with a list of addresses and then, once satisfied with the results, give the series a macro name so that it can be used over and over.

magnetic disk   A hard or floppy disk that holds magnetized data.

magnetic head   The mechanism that reads and writes data on a magnetic disk.

magnetic ink   An ink that contains chemicals that can be detected, and read, by special sensors. Magnetic ink is used for security purposes in industries such as finance and banking.

magnetic ink character recognition   The process of recognizing characters printed in magnetic ink.

magnetic ink reader   The device that reads characters written in magnetic ink.

magnetic storage   Storing data on disk and tape using the principles of magnetism.

magnetic tape   Plastic tape with a magnetic surface that can be used for storing data; often associated with storing data from mainframe computers, although also available for use with personal computers.

magnetic tape cartridge   A magnetic data storage tape housed in a cartridge-shaped container.

magnetic tape cassette   A magnetic data storage tape housed in a cassette-shaped container.

magnetic tape unit   A device used for reading and writing on magnetic tape with a magnetic tape head.

magstrip   A strip of magnetized plastic used for holding a small amount of data, often attached to the back of a credit card.

mail   In the computer industry, generally ELEC-TRONIC MAIL, which is the ability of users to send messages and other data from one computer to another across telecommunications lines.

mailbox   An area of memory that is assigned one
user to store any electronic mail messages sent
by other users.

mail-merge   A word processing function in which
one letter is personally addressed to many re-
cipients, often accompanied by personally addressed envelopes. While each word processing
program performs this function differently, a
mail-merge basically involves creating a letter
with the address area and personal greeting
coded with symbols that are recognized by the
program. A separate name and address file is
also created. When the mail-merge function is
run, the letter is printed repeatedly, each time
with a different greeting and address information from the name and address file.

mainframe   A large computer designed to handle
the information needs of many users in large
corporations. Mainframes are expensive and
complicated to support, often requiring special
air-conditioning and RAISED FLOORING, and,
while still heavily used, being supplemented
with, and even replaced by, smaller computers
and personal computers on local area networks.

main memory   Another term for MAIN STORAGE or
RAM. Main memory is the area of memory in
which reside the program commands and infor-

mation currently being used, as opposed to whatever is stored on floppy disks or the hard disk.

main storage (See MAIN MEMORY.)

maintenance The continual "care and feeding" of hardware or software. From a hardware perspective, can include anything from deleting unwanted files from the hard disk to cleaning dust from the interior of a computer. From a software perspective, includes activities such as installing new releases and testing them.

maintenance programmer A computer programmer whose major duties are maintenance-oriented. Generally, maintenance programmers work in large MIS departments.

maintenance routine The standard set of procedures in maintaining computer equipment or a software program.

malfunction A breakdown of a computer or peripheral device.

Management Information Systems (MIS) Used often in organizations to replace DATA PROCESSING. Refers to the overall corporate information system, including hardware, software, data, and the people responsible for developing and maintaining the system, and reflects the philosophy that the computer system exists to provide management with critical information, not simply to process data. The abbreviation is MIS.

management report A report generated by a software program containing information that is particularly relevant to management, such as usage statistics and related costs.

manual input Data loaded into the computer by the user, generally through the keyboard.

manual operation A procedure performed by a user rather than automatically by the computer system. For example, when a user deletes old files from a hard disk, one by one, this is a manual operation (software programs are available to perform this function).

map A "bird's eye view" of a computer program, including program elements and variables and where they are stored in the system. The map is especially useful if the program needs to be changed at some point in the future.

margin The blank space at the edges of a page of text. Word processing programs facilitate the creation of margins by making them easy to set and, with word wrap, to keep text within the margins.

mark   A symbol that indicates a function to be performed. For example, word processing programs use different marks to indicate areas in a document where an address or name will be inserted.

Massachusetts General Hospital Utility Multi-Programming System   A programming language designed specifically for the medical community for organizing hospital records.

massage   To modify, or process, information.

mass storage   A broad term for methods of storing large amounts of information. Mass storage devices include a hard disk, CD-ROM disk, and tape storage.

master–slave environment   An information system consisting of a central, or primary computer, that is attached to other computers and controls their operation.

material requirements planning   A process in manufacturing facilities for managing the inventory of parts and supplies needed to produce a final product. Increasingly, this process is computerized, often as part of a Computer-Aided Manufacturing system. The abbreviation is MRP.

math coprocessor   A hardware device installed in the CPU to enable it to perform special operations such as advanced computations.

matrix   A grid with lines that form rows and columns, such as used in a spreadsheet.

matrix printer   Another term for DOT-MATRIX PRINTER.

mature system   A computer system that is fully operational and, in some cases, at the point where it will not be able to handle any new demands.

MB   The abbreviation of MEGABYTE.

MBASIC   The abbreviation of MICROSOFT BASIC, a computer programming language for personal computers.

Mbyte   Another means of referring to MEGABYTE.

MCA   The abbreviation of MICRO CHANNEL ARCHITECTURE, an architecture developed by IBM for use in connecting PS/2 personal computers with each other, and with other devices.

MCGA The abbreviation of MICROCOLOR/GRAPHICS ARRAY, a graphics system used with IBM PS/2 personal computers.

MDA The abbreviation of MONOCHROME DISPLAY ADAPTER, a video standard for monochrome monitors used with IBM and compatible personal computers.

mean time between failures A rating used in evaluating the reliability of hard drives. Abbreviated MTBF, it is measured in hours, and the average hard disk should operate 40,000 hours without a major failure.

media Devices that can be used for storing data, including tapes, optical disks, and floppy disks.

medium The singular form of MEDIA.

meg A shortened version of MEGABYTE.

mega One million, used as a prefix.

megabyte One million bytes.

megacycle One million cycles per second.

megaflop One million floating-point operations per second, a measurement used in determining computer speeds.

megahertz One million cycles per second, as it relates to electrical transmission; used as a measurement of personal computer processing speeds.

memory A generic term for storage areas in the computer. Usually, refers to storage of data on a chip, as in main memory.

memory allocation The process of storing data and programs to specific areas of memory.

memory board An expansion board that adds additional memory to a computer.

memory chip A computer chip attached to a memory board for storing data and programs.

memory dump A printout that lists everything stored in a computer's memory.

memory management A category of software tools and techniques used in monitoring and optimizing a computer's memory.

memory resident   Programs permanently stored in the computer's memory rather than on a storage device. Generally, memory-resident programs are small applications such as calendars installed by the user.

menu   A list of options from which a user can choose, or a list of programs that a user can access. Within a program, additional options can also be listed on a menu.

menu bar   A menu arranged in the form of a horizontal line across the computer screen. Generally, each item on a menu bar is connected to another menu that appears once the user chooses it.

menu-driven software   Software that is designed so that commands and other functions are chosen from a series of menus.

merge   To combine two separate files in a way that, in the resulting file, the original organization of neither file is changed. For example, merging two alphabetically ordered name files will result in an alphabetically ordered file.

message   A shortened term for ELECTRONIC MESSAGE, a communication sent from one user to another.

message retrieval   The ability of the recipient to locate a message after it has been sent electronically.

method   A standard approach to a task.

metric system   The system of measurement used in Europe and in many other areas of the world.

MFLOP   The abbreviation of MEGAFLOP.

MFP   The abbreviation of multifunction peripheral, a device that performs functions normally requiring multiple devices. Examples of MFPs include a fax/modem and a laser printer/photocopy machine.

MHz   The abbreviation of MEGAHERTZ.

MICR   The abbreviation of MAGNETIC INK CHARACTER RECOGNITION.

Micro Channel architecture   (See MCA.)

microcolor/graphics array   (See MCGA.)

Microcom Networking Protocol   A communications protocol for high speed modems, developed by Microcom, Inc.

microcomputer   Used interchangeably with PER-SONAL COMPUTER. Microcomputers are the smallest category of computer and also the easiest to use and least expensive.

microcomputer applications   A generic term for computer applications in settings such as the home, education, and business, that involve the use of microcomputers.

microcomputer system   A microcomputer-based information system, including the CPU, application software, and peripherals such as a modem and printer.

microfiche   A small sheet of film on which the miniaturized reproductions of pages from a book or other publication are stored. Special devices read the information from a microfiche.

microfilm   A roll of film on which miniaturized pages of information are stored, similar to a microfiche.

microfloppy disk   Another term for floppy disk, though assumed to refer to the 3½-inch size.

microform   A generic term for the category of information storage that includes microfilm and microfiche.

microjustification A more finely tuned justification provided by some word processing programs, with narrow spaces added between the letters on a line of text so that the line becomes more attractive and readable.

micron One millionth of a meter.

microphone In multimedia applications, an input device on a personal computer that is able to record human sound, such as attaching voice messages to a word processing document or memo.

microprocessing unit Another term for MICROPROCESSOR.

microprocessor Another term for CPU. A microprocessor is a silicon chip that contains the personal computer's central processing unit.

Microsoft Windows A widely used graphical user interface for DOS-based IBM and compatible personal computers. Microsoft Windows makes extensive use of icons for functions such as choosing applications and issuing commands, which the user selects with a mouse.

microspacing The spacing process microjustification, in which the computer adds narrow spaces between some letters and words to make them more readable.

MICR reader A device that reads information printed in magnetic ink.

MIDI The acronym for MUSICAL INSTRUMENT DIGITAL INTERFACE, a standard for using the computer to produce electronic music; provides a means for computers to work with a synthesizer, and for music programs to be used for developing musical scores.

mini A slang term for MINICOMPUTER.

minicomputer A computer that, in size and power, lies somewhere between a personal computer and a large mainframe. Minicomputers are used in smaller organizations, which often do not need the power of a large mainframe, or in individual departments of a large organization.

minifloppy Generally, a 5¼-inch floppy disk.

MIPS The acronym for MILLIONS OF INSTRUCTION PER SECOND, used in measuring a computer's speed. MIPS refers to the average number of instructions that a computer can execute in one second. Understanding a computer's average MIPS can be helpful in making decisions about which computer to purchase; however, MIPS can also be misleading, because the computer is busier and slower, for example, during the day than at night.

mirroring In data storage on mainframe computers, data stored twice, on two separate sets of storage devices. This offers protection in case of a system failure that destroys one set of data.

MIS The abbreviation of MANAGEMENT INFORMATION SYSTEM.

MNP The abbreviation of MICROCOM NETWORKING PROTOCOL.

mode The current condition of a program. For example, a word processing program can be in insert mode or typeover mode.

model A computer-generated prototype of a system that can then be used as a basis for changing the original. For example, a model of a factory can be created on the computer, and various solutions to a production problem can be tested.

modeling The process of creating and testing computer models.

modem A device used to connect a personal computer with a telephone line, so that the computer can be used for accessing online information or communicating with other computers.

modify To change or alter a program.

modular design A program or computer system created in separate pieces, or components, that can be used separately or together in various ways. Modular design provides a basis for creating unique systems and changing them as necessary.

modular programming Designing a computer program in modules that can be reused and recombined to create new functions.

module A piece or component of a modular design.

monitor A video display for a computer.

monochrome A single color, as opposed to multiple colors.

monochrome display adapter (See MDA.)

monochrome monitor A monitor that offers only one color, generally amber or green. Monochrome monitors are generally used for applications such as spreadsheet and word processing in which color is not essential.

monospacing　A font in which characters all have the same width.

motherboard　Another term for SYSTEM BOARD, the main circuit board of a microcomputer. The motherboard contains the CPU, ports, and memory.

Motorola Microprocessors　A family of microprocessors manufactured by the Motorola Corporation and used in Macintosh personal computers.

mouse　A device used to move the cursor around the screen, to select functions to be performed, and to issue commands. A mouse has two buttons that are clicked to make selections.

mouse button　The button on the mouse used to make selections and issue commands.

mouse pointer　A small arrow on the computer screen, similar in function to a cursor, that moves whenever the mouse moves.

moving-bar menu　A menu in which selections are made by moving a highlighted, or colored, bar from one option to another. The bar usually appears over an option. Generally, once the desired selection is highlighted, it is selected by pressing the Enter key or clicking the mouse button.

MRP   The abbreviation of MATERIAL REQUIREMENTS PLANNING.

ms   An abbreviation of MILLISECOND.

MS-DOS   The abbreviation of MICROSOFT DOS.

msec   An abbreviation of MILLISECOND.

MS-Windows   The abbreviation of MICROSOFT WINDOWS.

MTBF   The abbreviation of MEAN TIME BETWEEN FAILURES.

multiaccess computer   Generally, a midrange or mainframe computer that can be accessed by multiple users as they run applications and retrieve data.

multicolor/graphics array   (See MCGA.)

multicomputer system   A computer system that uses more than one CPU.

multifrequency monitor A video monitor that can support a range of video standards, such as CGA or VGA, as opposed to only one standard.

multifunction board An expansion board that adds more than one new capability to a personal computer.

multifunction peripheral (See MFP.)

multimedia A generic term for computer applications that combine standard computer capabilities with other media such as video and sound.

multiple-user system A computer system designed to meet the information needs of more than one user simultaneously.

multiplex To send more than one message, over the same transmission line, simultaneously.

multiplexer A device that provides a means of transmitting more than one message simultaneously.

multiprocessing The ability of a computer to perform multiple operations at the same time. For example, a large mainframe computer can run multiple applications and serve many users at the same time. Increasingly, powerful personal computers are also able to run more than one program simultaneously.

multiprogramming Similar to multiprocessing; the ability of a computer to run two or more programs at the same time.

multiscanning monitor Another term for MULTI-FREQUENCY MONITOR.

multisync monitor Another term for MULTIFRE-QUENCY MONITOR.

multisystem network A network with more than one CPU attached, such as two mainframe computers, or one mainframe computer and one or more local area networks, for example. Users may have a choice in terms of which system they access.

multitasking Another term for MULTIPROCESSING.

multiuser system A system in which more than one user can access the computer simultaneously. Examples include a mainframe computer with many terminals attached, or a local area network with more than one personal computer attached.

multivolume file   A large file that cannot be stored on a single medium but must occupy, for example, two or more floppy disks.

MUMPS   The acronym for MASSACHUSETTS GENERAL HOSPITAL UTILITY MULTI-PROGRAMMING SYSTEM.

Musical Instrument Digital Interface   (See MIDI.)

musical language   A computer language that can be used to translate musical notes into a code usable by the computer.

music synthesizer   A device that plays electronic music and can be attached to a computer to play music developed with a musical language.

MUX   A slang term for MULTIPLEXER.

mylar   A coating often used to coat floppy disks. Mylar was developed by Dupont.

N

name An identifier for a file, data field, or program, comprised of a series of characters that has meaning to the user. Generally, a name can only include up to eight alphanumeric characters, though this varies among programs.

nano A prefix that means one-billionth.

nanosecond One billionth of a second; used in measuring computer processing speeds.

narrowband Data transmission limited to small amounts of data.

National Bureau of Standards The government agency responsible for creating and enforcing technical standards for computers.

National Computer Graphics Association A professional association of individuals involved in computer graphics.

National Television Standards Committee A standards committee responsible for setting technical standards for video and television.

natural language A language, like English, used by human beings. Computer languages such as fourth-generation languages, which approximate the way in which humans speak, are often referred to as natural languages.

NBS See NATIONAL BUREAU OF STANDARDS.

NCGA See NATIONAL COMPUTER GRAPHICS ASSOCIATION.

NDBMS The abbreviation for NETWORK DATABASE MANAGEMENT SYSTEM, a system for managing data stored on multiple CPUs, all connected by a network.

near letter quality The highest-quality print available from a dot-matrix printer, with print resolution close to that available from a standard typewriter.

NetBIOS   A program used on DOS-based IBM PCs to add additional local area network capabilities.

NetWare   A major local area network operating system developed and marketed by the Novell Corporation.

network   An information system based on two or more computers connected through telecommunications hardware and software. May include a combination such as personal computers connected on a local-area network, mainframes in separate geographical areas connected over telephone lines, or some combination.

network database management system   (See NDBMS.)

network interface card   A printed circuit board inserted into a computer and used to connect it to a network.

NIC   The abbreviation for NETWORK INTERFACE CARD.

NiCad battery pack   Nickel-cadmium batteries, often used to provide battery power to laptop computers. Nickel-cadmium batteries are relatively powerful but run down after approximately three hours and must be recharged.

NLQ   The abbreviation of NEAR LETTER QUALITY.

node   A point on a network where processing takes place; can refer to a computer or a device such as a printer.

noise   An unwanted disturbance on a telecommunications line, often resulting in disruption of data transmission.

nonimpact printer   A printer based on a technology such as ink jet or laser, which does not revolve around a hammer being struck on a ribbon. Nonimpact printers generally produce much higher-quality print with much less noise.

nonprocedural query language   A simplified language available to users for extracting information from a database; allows the user to focus on what information is needed without having to formulate the numerous steps for getting to it, as required with procedural languages.

nonswitched line   A permanent line connecting two computers.

nonvolatile memory   Storage that holds its contents even when power is turned off. ROM, or read-only memory, is an example of nonvolatile memory.

NOR operator An operator used in Boolean logic that is true only if the statements before and after the NOR are both false.

notebook computer A small portable computer, about the size of a standard notebook, and weighing approximately five pounds.

NOT operator An operator used in Boolean logic that is true if the statement after the NOT is false, and false if the statement after the NOT is true.

ns The abbreviation of NANOSECOND.

NSFnet A wide-area network, developed by the National Science Foundation, linking government agencies, universities, and various research organizations.

NTSC See NATIONAL TELEVISION STANDARDS COMMITTEE.

NuBus The expansion bus used with most Macintosh computers.

null In the computer industry, empty or no value. Zero is a numerical value.

null character The character used to imply no value. Null characters vary among programs, though a space is often used in personal computer applications.

null-modem cable   A cable used to connect a personal computer directly to a larger computer through their communications ports.

number cruncher   A computer or a program capable of performing mathematical computations at a rapid rate.

numeral   The standard means of representing a number, such as 7 and 234.

numeric character   Another term for NUMBER or DIGIT.

numeric coding   Computer programming that uses numerals, instead of alphabetic characters, to write program code.

numeric coprocessor   Another term for MATH CO-PROCESSOR.

numeric data   Data that include only numbers.

numeric keypad   A separate set of number keys, including a decimal point key, provided on more sophisticated (101-key) computer keyboards. The numeric keypad is activated by pressing the

Num Lock key, and can be used like the keypad of a calculator for more rapid keying of numbers.

numeric system   A system for representing numerals, such as binary or hexadecimal.

Num Lock key   The key associated with a numeric keypad. When pressed, the numeric keypad is activated. Otherwise, the keys of the numeric keypad act as cursor control keys.

◊

OA The abbreviation of OFFICE AUTOMATION, the used of computers and sophisticated information management technology in an office setting. Word processing, spreadsheet, and database management are examples of office automation applications.

object code Program code that has been run through a compiler. Program code that has not yet been is referred to as SOURCE CODE.

object-oriented programming An approach to programming in which programs are developed in modules, or objects, that can be combined or modified to create new capabilities and programs.

object-oriented programming language A programming language used in creating object-oriented programs, such as C++ and Smalltalk.

object program A program that has been run through a compiler.

OCR The abbreviation of OPTICAL CHARACTER RECOGNITION, a technology for transferring information from a printed page, such as book or magazine, directly into a computer. This information is then translated into a standard format understood by the computer so that it can be edited the same as other data.

octal A number system based on eight characters and, because they are easier to read than binary numbers, often used instead of the binary format.

odd header A header, generated through a word processing program, that appears only on the odd-numbered pages.

OEM The abbreviation of ORIGINAL EQUIPMENT MANUFACTURER, a hardware company that purchases parts and components from other companies and uses them to build a product that it subsequently sells under its own name. Many personal computer manufacturers buy many of the components of their products from other companies.

office automation   (See OA.)

offline   Peripherals, such as printers, that are not currently connected to the CPU.

offset   In desktop publishing, moving the contents of a page further to the right to allow extra space for binding.

on-board   A memory chip, or a device such as a modem, that is attached to a circuit board.

on-board modem   A modem attached to an expansion board.

online   Describes peripherals that are directly connected to a CPU and available for use; sometimes used to describe a CPU that is turned on and ready for use.

online database   A database of information available for users to access by modem over a telephone line. Examples include summaries of newspaper and magazine articles and legal documents.

online fault tolerant system   A computer system designed to continue operation even after a disaster.

online information service A subscription service that, for a fee, provides users with access to various online databases and related services. CompuServe is an example.

OOP The abbreviation of OBJECT-ORIENTED PROGRAMMING.

OOPL The abbreviation of OBJECT-ORIENTED PROGRAMMING LANGUAGE.

open The function performed by the operating system, generally "behind the scenes," after the user issues a retrieval command, before the user can retrieve data from a database.

open architecture A computer system based on standards readily available for any vendor to use in creating products. With open architecture, companies can purchase hardware and software products and integrate them into their network with no concern about one product being compatible with another.

open file A file that can be accessed by users for retrieving or manipulating data.

Open Systems Interconnection Model A set of standards for communication between computers and related devices developed by the International Standards Organization.

operand A piece of data manipulated in a computer program.

operating environment The commands and capabilities that users have available to them as a result of using a specific operating system. For example, DOS users have available to them a wide range of DOS commands. These commands constitute the DOS operating environment.

operating system The software program that controls the operating of the computer and acts as an interface between the actual machine and the software applications installed by the user. The operating system creates and organizes files and directories and controls the operation of devices like printers. Examples of operating systems are MS-DOS and OS/2.

operation Any specific action performed by a computer; for example, when a computer adds a column of numbers or sends a document to the printer.

operations A function of the MIS or data processing department concerned with the day-to-day maintenance of the computer and related hardware.

operations management An area of computer science focused on making decisions about the organization's computer hardware.

operator A symbol or word used to represent an operation, like a mathematical operator or a Boolean operator.

optical character reader A device that can read characters from a printed page as input to the computer.

optical character recognition (See OCR.)

optical disk A storage medium that can contain large amounts of data. The data are written with laser technology. The most commonly used optical disk is CD-ROM.

optical engineering A branch of engineering concerned with the use of light as a basis for telecommunications.

optical fiber A glass thread that serves as a basis for high-speed, and reliable, data transmission. Optical fibers are being increasingly used both in data and voice communications.

optical laser disk Another term for OPTICAL DISK.

optical reader wand A penlike device used to read bar codes.

optical scanner An optical character reader that scans through a printed page.

optimization The process of designing, and refining, a computer system or application until it functions at its highest potential.

optimum programming Creating an application program with an underlying goal such as efficient use of computer memory or rapid processing time.

option A choice on a menu or in a dialog box.

Option key On a Macintosh keyboard, used with other keys to generate alternate characters, and commands, depending on the software program. (For IBM and compatible keyboards, see ALT KEY.)

order To arrange data, or a series of commands, based on predefined rules.

original equipment manufacturer (See OEM.)

OR operator A Boolean operator that is true if the statements on both sides of the OR are true.

orphan A word processing term that refers to the first line of a paragraph, or a subheading, that appears as the last line of a page.

OS The abbreviation for OPERATING SYSTEM.

OSI The abbreviation for OPEN SYSTEMS INTERCONNECTION.

OS/2 The operating system generally used with IBM's PS/2 line of personal computers. OS/2 was developed by Microsoft Corporation and IBM, and provides features such as increased main memory over standard DOS as well as a graphical user interface.

output Virtually any type of data generated by a computer, including whatever appears on the screen or is printed or sent across a telecommunications line; can also include voice and sound.

output area The section of a computer's storage where data are held for output.

output buffer A holding area where output data are temporarily stored while waiting to be sent to an output device. For example, a large document waiting to be printed may be held in a buffer area while the printer processes it, rather than tying up the computer any further, so that the user can continue with other tasks.

output data   Data that have been processed and are ready for output, or have already been output, to a screen, printer, or other output device.

output device   Any device that accepts output from a computer, such as a monitor, modem, or printer.

output media   Physical media on which data can be output, such as a floppy disk or paper.

output stream   A line of output data, often in files, on its way to an output device.

outputting   The process of transferring data from the computer to an output device.

overflow condition   A condition that can occur during an arithmetic operation as a result of the computer attempting to process a number that is beyond its capacity.

overflow error   Another term for OVERFLOW CONDITION.

overhead   Generally, the operating system and any additional utilities, such as calendar software, that the user may have added. These programs, while necessary and useful, do not directly serve the needs of users in the same way as applications programs like spreadsheets.

overlaid windows   In a graphical user interface, windows that appear to be stacked like sheets of paper, each window identified by the types of icons that appear on it. The user clicks on one of the windows to bring it to the top of the stack.

overlap   When applied to the operations of a computer, the performance of multiple operations at the same time, such as editing a document while another document is being output to a printer.

overlapping windows   Another term for OVERLAID WINDOWS.

overprint   The means by which dot-matrix printers produce bold print, with the characters of specific words struck more than once to produce darker type.

override   To temporarily change a standard setting in a program for a specific operation. An example of an override is changing the standard margins in a word processing program for the purpose of printing one document, and then allowing the margins to return to the standard setting after the document is printed.

overrun   The result of sending too much output to an output device, for example, sending a large document to a printer with a small buffer and then leaving the program before the printing is complete. This is an error condition.

overstrike   Another term for OVERPRINT.

overwrite   To make a change in an already existing file and then replace the old file with the new version. For example, creating a document, saving it, retrieving it and making further changes, and then saving it again with the same document name.

overwrite mode   Another term for OVERPRINT.

P

pack To compress small files, by means of a software program so that they fit into a single storage area and require less memory. Packed files can be unpacked for later use.

package A software product that can be purchased and put to immediate use.

packaged software Another term for PACKAGE.

packed file A file that has been compressed to fit into a smaller storage space.

packet A chunk of data being sent over a telecommunications line.

packet switching A method of sending data in chunks, or packets, with a degree of reliability, over a telecommunications line. The line is dedicated to each specific packet while it is being sent and, afterwards, becomes available for the transmission of another packet.

packing The process of compressing a data file so that it requires less storage space.

pad character A character used in filling a field, such as a database field, when the actual data in it contain less characters than required. For example, if a nine-digit code is entered in a field that has twelve spaces, three additional pad characters might be added to fill the field. Not all programs require the use of pad characters, and the characters used differ among programs.

padding The process of adding PAD CHARACTERS to a field.

page In word processing, a page of text in a document.

page break The end of a page of text, occurring either when the number of available lines on the page has been reached or when the user chooses to end the page.

page description language A language in desktop publishing to define the format and contents, including print fonts, of a printed page. Adobe PostScript is an example of a page description language.

Page Down key The key abbreviated PGDN that, in a word processing program, moves the cursor forward a page at a time through a document.

Page Eject A command to a printer to advance a page forward, one page at a time.

page layout program A program to make text and graphics work together on a page, such as might be used in publishing a newspaper.

page preview A function provided in word processing programs that allows the user to view each page of a document on the screen before actually sending it to be printed. This way, changes can be made without wasting time and paper.

page printer A printer that composes each page internally and prints it as an entity rather than one line at a time.

page reader An optical scanning device that scans a whole page and translates it into a code usable by the computer.

page skip A command inserted into a document that, when reached during the printing process, signals the printer to begin a new page.

pages per hour The number of pages a printer can print in an average hour, used to measure printer speed.

pages per minute The number of pages a printer can print in an average minute, used to measure printer speed.

Page Up key The key abbreviated PGUP that, in a word processing program, moves the cursor backward a page at a time through a document.

pagination Designating the way in which the pages in a document should be numbered, including the location of the page number and whether a number should appear on each page, or only on even- or odd-numbered pages.

paging Retrieving a page from a document and viewing it on the screen of the monitor.

paint To fill graphic images with color. Also referred to as FILL.

paintbrush Tools available in a paint program that provide the user with various options for filling in graphic images with color.

painting Using a paintbrush to fill a graphic image with color.

paint program A program to draw various images on a monitor screen and fill them with color.

palette The group of colors available to the user of a paint program.

palmtop A computer small enough to fit in the palm of an adult. Palmtop computers are used for performing calculations and for specialized applications like foreign language dictionaries and address books.

panel Another term for CONTROL PANEL, the physical area of a computer on which control buttons and switches are located.

paper feed The means by which paper is advanced through a printer. For example, a dot-matrix printer uses perforated paper and a tractor feed.

paradigm An idea, or a concept, that may lead to a technological advancement.

parallel The ability to simultaneously process more than one element of data. For example, printers are often connected to a parallel interface, which speeds the printing process.

parallel computer A computer designed so that data and applications are duplicated, in two separate areas of the machine. As a result, applications are protected in case of a disaster.

parallel interface A piece of equipment, such as a port on a computer, designed to simultaneously process more than one element of data.

parallel port A port designed to provide a parallel interface, often used for a printer.

parallel processing Processing data and applications through the use of two or more CPUs. One CPU may be used for processing one part of a program, and another CPU used for another part, to provide faster processing.

parameter An attribute of a computer program. For example, margins and type sizes are parameters in a word processing program.

parent directory Directories are generally organized in a tree structure, and a directory's parent is the one that resides a level above it in the tree. Subdirectories are all related to a parent directory.

parity checking In telecommunications, the process of assuring that a message has been accurately and reliably sent by checking for transmission errors.

park To lock a personal computer's floppy disk drive when transporting it to make sure that it is not jarred, and possibly damaged, during the move. Parking is especially important with portable and laptop computers, which are often moved.

partition To divide main memory into pieces, with each assigned to a specific program. The result is more efficient use of memory and faster processing.

Pascal A high-level programming language in structured programming, most often in the academic world. It was named for Blaise Pascal.

pass Going completely through a set of data, which is what a computer program does as it performs a specific operation on the data and then moves onto the next instruction.

password A secret word that a user enters to gain access to the computer system, to a particular application, or to a file within an application. Passwords are used for security purposes, to keep unauthorized users from private data. A password may be assigned by the organization or generated through the application, or the user may choose his or her own password.

paste To move a section of a document, like a paragraph, from one location to another in the same or a separate document.

patch A piece of program code provided by the vendor, and added by the user, to repair an error, or BUG, in the program. The patch may be provided over the telephone or sent to the user on a floppy disk or tape.

path The route taken by the operating system as it goes through directories to locate a file specified by the user.

Pause key A key that, when pressed, will stop the flow of data across a display screen; used when the system is scrolling data too rapidly for the user to read.

PC The abbreviation of PERSONAL COMPUTER.

PCB The abbreviation of PRINTED CIRCUIT BOARD, the board inside the computer on which electronic components such as chips are installed; the foundation of any computer.

PC-DOS Another way of referring to Microsoft's MS-DOS operating system, generally used by IBM.

PC FAX Another term for FAX MODEM.

PCL The abbreviation of PRINTER CONTROL LANGUAGE, Hewlett-Packard's page description language.

PCM The abbreviation of PLUG-COMPATIBLE MANUFACTURER, a computer hardware manufacturer that makes equipment that can be used interchangeably with products developed by other vendors. Plug-compatible vendors often offer their products at a lower price.

PCX A standard graphics file format developed by ZSOFT and generally used with IBM and compatible personal computers.

PDL The abbreviation of PAGE DESCRIPTION LANGUAGE.

peek A function that allows the user to look at the contents of a file without having to first retrieve them into main memory.

peer-to-peer architecture A system design in which the CPUs connected in a network are all on the same communications level. For example, rather than a mainframe passing data and instructions to a local area network, in a peer-to-peer arrangement both the mainframe and local area network are equally capable of sending and receiving data and instructions.

pel The abbreviation of PIXEL.

penetration Unauthorized access to computer system.

perform The carrying out of instructions by a computer.

performance monitor A software product that measures the computer's ability to carry out instructions, analyzing factors such as speed and reliability. Performance monitors are generally used with mainframe and midrange computers, though they are also available for local area networks.

peripheral equipment The devices connected to a computer to perform additional functions for users. These include additional storage devices, like CD-ROM, modems, and printers.

permanent storage Another term for STORAGE.

personal computer Another term for a MICRO-COMPUTER, based on its ability to serve all of the computing needs of an individual user.

personal identification number (PIN) A number assigned to a user to serve as a security password to gain access to the system. The most common use of personal identification numbers is with automatic cash machines.

personal information manager A software program that enables the computer to manage day-to-day organization details, with features such as a calendar, a note and list maker, and a calculator.

PgDn key The abbreviation for PAGE DOWN KEY.

PgUp key The abbreviation for PAGE UP KEY.

Phoenix BIOS A BIOS developed by Phoenix Corporation that duplicates the actions performed by IBM's BIOS.

physical security Computer security that employs locked doors and special keys for accessing the system, as opposed to passwords.

PIC A slang term for LOTUS PICTURE FILE, a graphics file format developed by Lotus Corporation.

PICT file format   A graphics file format developed
by Apple.

picture processing   The use of image processing
technology for storing and retrieving pictures.

PIM   The abbreviation of PERSONAL INFORMATION
MANAGER.

PIN   The abbreviation of PERSONAL IDENTIFICATION
NUMBER.

pin   The tiny pin-shaped mechanisms installed on
the hammer device of a dot-matrix printer that
hit the ribbon to create type on a sheet of paper.
Resolution with a dot-matrix printer is best
when the number of pins is twenty-four.

pin feed   A means of feeding paper through a
printer, in which each side of the sheet of paper
has a perforated strip with holes in it. The paper
is pulled through the printer by means of a wheel
on each side of the printer, with nail-shaped
devices that go through the holes.

piracy   The theft of a software product when a user
passes on his or her copy to another user. Piracy
prevents the software vendor from collecting
due compensation, and also causes a decline in

revenue that the vendor would spend on further development and product support.

pitch   The number of characters contained in a line of text. For example, 12-pitch type means twelve characters printed per inch.

pixel   The abbreviation of PICTURE ELEMENT, a dot on a display screen used in creating graphic images. Millions of pixels are combined to create a graphic screen image, with the clearest images resulting from the largest number of pixels per inch.

planar board   Another term for MOTHERBOARD.

plansheet   Another term for SPREADSHEET.

plasma display   A flat-panel display technology based on the use of neon. Plasma displays are often used with laptop computers.

platen   The roller around which the paper revolves in a dot-matrix printer.

platform   The operating system or hardware foundation of a computer system. For example, an IBM mainframe is an example of a platform, as is an Apple Macintosh personal computer.

platter The round disk portion of a hard disk where the actual data are stored.

PL/1 The abbreviation of PROGRAMMING LANGUAGE ONE, a computer programming language used with mainframe computers.

plot To draw a diagram with the use of a plotter.

plotter A device that can be attached to a computer and draw graphs and pictures. Plotters use pens and thus create drawings based on lines.

plug The connector at the end of a cable that links one device to another.

plug compatible A piece of equipment made by one vendor that can be used directly with, or even replace, that of another vendor.

plug compatible manufacturer (See PCM.)

pocket computer Similar to a PALMTOP COMPUTER.

point To select a function by moving the pointer, an arrow that appears on the screen, by means of a mouse, until it rests on the appropriate icon.

pointer The arrow that appears on the screen in a graphical user interface, and that is moved by mouse.

pointing device   A device, such as a mouse or a joystick, that moves the pointer around the screen.

point-of-sale terminal   A device in retail establishments that performs the function of a cash register but is also connected to a computer. As a result of this direct connection, sales trends and stock can be monitored on a more immediate basis.

polling   A process by which a communications device continually questions other devices about whether or not they have data to transmit. A modem is the most common device to use polling.

pop-up utility   A small program, like a calendar, stored in main memory and always available for use, even when another program is currently running.

pop-up window   A window displayed when the user chooses a function. Pop-windows have a limited number of choices and, once the user clicks on one of them, the pop-up window disappears.

port   To move an application from one platform to another, such as moving a mainframe application to a local area network.

portable computer   A personal computer that can easily be moved from one location to another. Portable computers are relatively lightweight, enclosed in durable cases, and can usually be powered by battery. The portable computer family includes laptop, notebook, and palmtop computers.

portable software   An application designed so that it can be moved from one platform to another.

portrait   In word processing, a page that is taller than it is wide, as opposed to landscape, which is the opposite. Standard documents are printed in portrait mode.

PostScript   The widely used page description language, developed by Adobe Systems, used in laser printing documents in desktop publishing applications. PostScript provides users with a wide range of font options and graphics capabilities.

power down   To turn off a computer or other device.

power on   To turn on a computer or other device.

power supply   A mechanism installed in a computer that converts the AC current from the electrical outlet into the DC current required to run a computer.

power surge A sudden burst in the electrical stream that supplies power to a computer, often resulting in loss of unsaved data and, in some cases, equipment damage.

pph The abbreviation of PAGES PER HOUR.

ppm The abbreviation of PAGES PER MINUTE.

presentation graphics Graphics used in creating presentations, including charts and bold graphics, and often used in a business setting.

preset Another term for INITIALIZE.

press To push or depress a key or button.

preventive maintenance The process of cleaning and replacing worn parts on computers and related equipment to avoid an equipment failure.

preview To view a document onscreen before actually sending it to the printer; a function provided in most word processing programs.

primary storage Another term for MAIN STORAGE.

print control character   A character inserted into a document to control certain printer operations. An example of a print control character is one that causes the printer to start a new page at various points in the document. Print control characters are a word processing program function.

print density   Similar in meaning to PITCH, refers to the number of characters within any unit of measurement, including characters per line and characters per page.

printed circuit board   (See PCB.)

print element   The mechanism in an impact printer that transfers the type to paper. A daisy wheel is an example of a print element.

printer   A device that transfers words or images onto paper.

print head   Another term for PRINT ELEMENT.

printout   Any information stored in a computer that has been printed on paper. Examples include a listing of the contents of a database on standard computer paper, as well as a carefully formatted document on letterhead. Also referred to as HARD COPY.

print quality The overall condition of printed material, taking into account factors such as clarity of type and spacing between letters and words.

Print Screen key A key provided on some keyboards that sends the current contents of the screen to a printer and prints it on paper; generally used when the user wants a quick record of the results of a command or an error message.

print server A computer attached to a network with the sole function of managing one or more printers. For example, local area networks often include a personal computer that serves as a print server. Users send their documents to the print server, which, in turn, chooses the appropriate printer for the document and monitors the execution. In the meantime, users can go on with other work while the print job is processed.

print spooling The process of sending documents off to a storage area on a disk, called a BUFFER, where they will remain until the printer is ready for each one in turn.

private line A communications line provided for the sole use of one user.

privileged instruction A command available for use only by the operating system and not by individual application programs. This assures that application and operating system commands are not confused with each other.

problem analysis The first step in developing a computer program, with users and technical professionals discussing the problems that the program will need to solve.

problem description A formalized summary of the results of the problem analysis, agreed upon by all parties involved, with a strategy about how it will be solved.

procedure A set of steps that an application program goes through in executing a command from a user or otherwise performing an operation.

process What a computer does when it accepts data from a user, responds to one or more commands, and produces the desired response.

process conversion Focusing on one of the processes performed by the computer and changing it so that it occurs in another manner.

processor Another term for CENTRAL PROCESSING UNIT.

processor bound Time-consuming processes, such as searches of large databases, that are slowed down because of the time required by the CPU for completion.

production The work performed by the computer for users on a day-to-day basis.

production run The first time that a new program is used on a day-to-day basis. Generally, programs are tested and "debugged" before they are used in production work.

program A list of instructions, written in a computer language, that will cause the computer to process data and perform other specific operations. A broad term, it can refer to those that are created by programmers as well as by users. Programs can be written for specific applications, like financial management or word processing, or to control the operating system, like DOS, and are also created to manage data, as in database management systems.

program coding The actual process of writing a program, using a programming language in a code understood by the computer.

program development cycle The steps the technical staff goes through in creating a program, including defining the problem, deciding on the best approach, writing the actual program code, testing it, and then making any additional changes before it can be used for production work.

program file A storage file that holds computer programs.

program flowchart A diagram that illustrates how a computer program works; includes lines that show the overall flow of the program, from one step to the next, with each major step in the program identified by a symbol. Especially useful to programmers as they begin the process of writing the program code.

program identification An identifier given to a program, including the program name and, often, a number identifying the current version of the program.

program language A language used in writing a computer program. Each program language has its own vocabulary, commands, and rules that must be adhered to if the program result is to be accurate. Program languages differ widely from each other in terms of the hardware they can be used with, the purposes they can be used for, and the amount of technical expertise required from the programmer.

program library A group of programs and sections of programs that may be modified for new purposes or reused in new programs, thus helping programmers avoid duplication of effort.

programmable function keys Keyboard keys whose function changes with each application program. For example, they perform different functions in a word processing program than in a spreadsheet program.

program maintenance The process of keeping a program updated as the needs of users, and organizational requirements, change. Also, program errors sometimes surface after a program has been in use for some time, and program maintenance also involves making repairs to prevent these errors.

programmer An individual who designs and writes computer programs.

programming The act of writing a computer program, from initial design to coding to final testing.

programming aids Software programs designed to assist programmers in developing computer programs. Examples include programs that assist with locating and fixing program errors.

programming language   Another term for PRO-
GRAM LANGUAGE.

program specifications   A formal document that
outlines how a program will be developed, what
hardware it will be written for, the data to be
accessed, and other related information.

program testing   A formal process that involves
testing all aspects of a program in simulated situations to make sure that it performs according
to specifications.

project   A large task with a specific goal, divided
into various steps to serve as milestones to measure progress toward completion.

project management software   Application software designed to aid in managing large projects.
Functions include assistance in defining the required steps for completion, setting schedules,
monitoring progress, and managing costs.

prompt   A message from the system to indicate that
it is ready to receive information or further instructions. Examples of prompts include a ques-
tion mark or a message such as "Please enter
command."

proofing program   A general category of software programs that can be used with word processing programs to check for spelling and grammatical errors. Many word processing programs have these features built in.

proportional font   A font in which the pitch, or width of each letter, reflects the actual width of the letter. This can be contrasted to a fixed-pitch font, in which each letter is of the same width. For example, in a proportional font, the width of the letter *m* is wider than the width of the letter *i*.

proportional pitch   Pitch, or width, that reflects the actual width of the letter, as used in a PROPORTIONAL FONT. Documents printed in proportional pitch are both clearer and more attractive to the human eye.

proportional spacing   Another term for PROPORTIONAL PITCH.

proprietary   A system design unique to a particular hardware or software vendor. Proprietary design limits users' freedom of choice, because they are locked into purchasing products that will work within one vendor's product line rather than being able to choose among products and services that best meet their needs.

protect   To institute security procedures that prevent unauthorized users from access. This might include attaching a password to a specific file, or requiring users to enter passwords and follow other security procedures for using the computer system.

protocol   A format, or a set of guidelines, for establishing communications between computers and other devices. Protocols are generally formulated by large governing bodies, such as the International Standards Organization, responsible for setting data communications standards.

PRT Sc key   The abbreviation of PRINT SCREEN KEY.

PS/2   The abbreviation of PERSONAL SYSTEM/2, the more sophisticated family of personal computers developed by IBM to follow its original family of IBM personal computers.

public carrier   A business such as AT&T or Sprint that sells communications capabilities through telephone lines.

public-domain software   Software products that are not copyrighted and are thus available to the general public to be copied and shared at will.

public network   A communications network that is available for access by anyone.

pull-down menu   A menu that appears when the user clicks on either an icon or a word listed in a title bar that is displayed across the top of the screen. The pull-down menu provides additional options and, once the user makes a choice, disappears.

punch card   In the early days of computing, input to a computer was provided through the use of small rectangular cards punched with codes representing program instructions as well as data. These cards were read by the computer in a similar, though much slower and inefficient, manner as the computer now responds to commands entered through the keyboard.

punctuation   The use of apostrophes, commas, periods, and other standard punctuation in computer programs to separate words and commands.

purge   To erase, or delete, data.

Q

QBE   The abbreviation of QUERY BY EXAMPLE, a simplified means of retrieving data offered by some database management systems. With QBE, the user is presented with a screen containing fields that the user can fill in to describe the data he or she is interested in retrieving. This is much easier than entering a series of commands, or using Boolean logic, to describe the needed data.

QIC   The abbreviation of QUARTER-INCH CARTRIDGE, the size of tape cartridge often used in tape backup systems for personal computers.

quality control   The process of testing and modifying computer programs during the development process to assure that the final product is as useful and bug-free as possible.

quarter-inch cartridge (See QIC.)

query To establish contact with a database management system and make a request for information.

query by example (See QBE.)

query language A special language used in obtaining information through a database management system, generally involving the use of relatively simple commands. Each database management system has its own query language.

question-answer Describes the process of interacting with the computer, with the user posing questions and the computer providing answers.

queue A line of files waiting for processing. For example, when a user sends one word processing document file after another to the printer, the files may remain in a queue waiting to be printed.

QWERTY keyboard The standard keyboard on typewriters as well as computer keyboards. The word *QWERTY* comes from first six keys, starting from the left, of the second row of keys. The QWERTY keyboard was designed to prevent rapid typists from jamming the keys. Alternate keyboards include the Dvorak keyboard, which is designed for faster typists.

R

race condition What happens when two operations are being processed simultaneously.

rack A metal chassis on which computer equipment
is mounted.

radio buttons Button-like images in a graphical
user interface that are used for "turning on"
certain features. When one button in a group
has been selected, the others cannot be used at
the same time.

ragged Text in a document that is not justified on
one margin or the other.

ragged left Text that is not justified with the left
margin, but aligned on the right.

ragged right   Text that is not justified with the right margin, but aligned on the left.

raised flooring   The floor in a computer room, raised to accommodate additional cabling and even water pipes to provide cooling for mainframe computers.

RAM   The abbreviation of RANDOM ACCESS MEMORY, the immediate, or "working," memory, that can be quickly accessed without requiring a search of storage areas. RAM is also referred to as main memory, and holds whatever application program, and associated data, is currently being used.

RAM cache   Extra-high-speed RAM available on some computers, utilizing static, rather than dynamic, memory technology.

RAM card   An expansion card containing RAM chips that can be plugged in like a standard expansion card and used to expand a computer's available RAM.

RAM disk   RAM that has been subdivided like a disk, for the duration of a computer session, to speed up a particular application. Subdividing RAM in this way is a temporary condition; when the computer is turned off, the RAM goes back to its original condition.

RAM resident Another term for MEMORY RESIDENT; programs that remain in main memory.

random access The ability to access data at random, regardless of the order in which they are stored.

random-access memory (See RAM.)

random files Files that are not organized in any set order.

random number A series of digits with no apparent logical order. Computers generate random numbers for research projects and contests, for example.

random number generator A computer program used in generating random numbers.

range A set of one or more contiguous rows or columns in a spreadsheet.

rank To arrange a set of data elements in either descending or ascending order.

raster graphics   The representation of graphic images in the form of bit maps, which are rows of closely spaced dots or pixels.

raw data   Data that may have been entered into the computer but have not been processed in any way to turn it into meaningful information.

RDBMS   The abbreviation of RELATIONAL DATABASE MANAGEMENT SYSTEM, a database management system based on the relational model. (See RELATIONAL MODEL.)

read   To retrieve information from a storage medium, such as a floppy disk, hard disk, or CD-ROM disk.

readme file   A file available in most application programs that contains additional information about the program, such as guidelines for using the program and instructions on how to obtain the documentation.

read-only memory   Computer memory containing programs that are critical to the operation of the computer, like the instructions necessary to boot the computer when it is turned on. Regardless of whether the computer is on or off, whatever has been stored in read-only memory remains.

readout Another term for PRINTOUT.

Read-Write Head The device that writes data on, and retrieves them from, a hard or floppy disk.

real storage Another term for MAIN STORAGE.

real time Events that happen in the here-and-now. Computers that operate in real time process instructions as they are entered rather than storing them for execution at a later time. For example, a bank that uses real-time processing with its automatic teller machines would update customer accounts as the transactions are made, rather than updating accounts at night and posting the changes on the next business day.

real-time clock A clock in the computer that keeps track of the actual time, even when the computer is not in operation.

reboot To restart a computer after it has gone down.

recalculate To compute the values in the cells of a spreadsheet after a change has been made.

receive To accept a message sent by telecommunications, such as through a modem.

receiver   The party who receives a message.

record   A set of various pieces of related information treated as one unit; often used in database management, but it has its roots in traditional information management. For example, a personnel record, which might be contained in a standard file cabinet or in an electronic database, contains information such as employee name, address, and social security number.

recording density   The amount of actual storage space available on a storage medium such as a floppy disk, measured in characters or bits per inch.

record layout   The format for arranging the information in a database record. This includes the length of each field on the record and the kinds of characters, alphabetic or numerical, that can be stored in it.

record length   The size of a database record, generally expressed in number of bytes or characters.

record locking   Preventing users from retrieving a record in a database while another user is modifying it. This helps to assure that users are accessing the most up-to-date version of a record.

record number An identifying number automatically assigned to each record through the database management system.

records management In an organization, responsibility for organizing and maintaining paperwork. The function lends itself to computerization, and it has been the focus of both hardware and software developments.

recover The ability of a software application, or a hardware device, to resume operations after experiencing an error.

recursive Another term for REPETITIVE.

red-green-blue monitor A high-resolution, full-color monitor.

redlining Marking edited text so that the next person to read it is aware of what has been edited. Many word processing programs including redlining features that automatically mark edited text, such as in bold text.

reduced instruction set computer A computer that recognizes a limited number of instructions—the ones performed most often—and because of this less-complex design, is much faster than more traditional, complex instruction set computers. Many of the newest computer work-stations are based on RISC technology.

reduction   Deleting unnecessary files to create us-able space on storage media.

redundancy   The concept of running two applica-tions programs at the same time, generally on two separate CPUs so that, in the event of a fail-ure on one CPU, the functions performed by the application, and the data, will not be lost.

reference manual   Documentation that accompan-ies a user manual, with explanations of how spe-cific functions are designed to work and, often, with definitions of error messages.

reformat   To change the way data are organized from one format to another.

refresh   To recharge a device or restore information that has started to disappear. Display screens need to be constantly refreshed with phosphor or the image begins to fade, and RAM also needs to be refreshed or data will disappear.

refresh circuitry   The electronic circuitry associ-ated with refreshing display screens or memory.

refreshing The process of continually recharging a
screen or memory.

refresh rate The rate at which the refresh action
is taken on a display screen. This varies by device and type of image displayed.

region An area in storage designated for a specific
purpose.

register A small storage area within the CPU that
holds data immediately before it is to be processed.

relational database A database with a design
based on the relational model.

relational database management system (See
RDBMS.)

relational expression A statement that expresses
the relationship between two or more variables,
using a relational operator.

relational model A data management model in
which data are stored in various tables. This
model offers great flexibility because users can
then establish relationships between various

types of data at will, for the purpose of a query, without being tied into predetermined formats. For example, a search for personnel information could span a range of tables, containing data such as employment history, address, and social security number, rather than being limited to the contents of a predefined personnel record.

relational operator   A symbol used in expressing relationships, such as *greater than* or *lesser than*.

relational structure   Another term for RELATIONAL MODEL.

release version   The most current version of a software product, containing the latest features.

reliability   The degree to which a computer, a device, or a software application can perform with a minimum of errors.

relocate   To move a computer program from one area of storage to another.

remark   A comment added by a programmer to a line of program code that explains why it was coded as such and what it is meant to accomplish. This is helpful for future programmers who may be responsible for maintaining the program.

remote A CPU or device physically located in a separate location yet connected on a network to another CPU or device; can refer to another area of the same building or a building in another city, state, or country.

remote access The ability of two separate CPUs or devices to communicate with each other over a network.

remote computing services Computing services, such as data processing, offered from a computer located at one location to users in other locations.

remote control The ability of one computer to access data from, and use applications on, another computer located at a separate site; for example, a user sitting in a hotel room and using a laptop computer and a modem to retrieve data from the corporate mainframe computer.

remote job entry Entering data in a computer, and initiating procedures, from terminals in one or more remote locations, as users on the factory floor of a large company might do.

remote processing Using computer applications at a remote location through terminals connected to a computer.

remote site Another term for a location that is REMOTE.

remote terminal A computer terminal that does no processing of information on its own but is connected to a computer in another location.

removable cartridge A storage disk that can be inserted into a special slot on a computer. Removable cartridges are similar to floppy disks in that they can be inserted and removed at will, yet they offer the durability and faster access of hard disks. The actual mechanism for reading and writing data is enclosed within the cartridge.

removable hard disk Similar in concept to a removable cartridge except that a removable hard disk does not have its own internal read/write mechanism.

report A computer-generated summary of data, often in an arrangement understandable to managers and users.

report generator A software program that aids the user in organizing information into useful reports. Often a report generator is included as part of a database management system.

reporting by exception A report that includes only exceptions to the norm rather than volumes of unnecessary routine details.

Report Program Generator A programming language for developing various business applications as well as creating reports.

report writer Another term for REPORT GENERATOR.

reproduce To copy information from one storage medium to another, such as from a hard drive to a floppy disk.

reprogramming Rewriting or otherwise modifying a program written for use on one computer system so that it can be run on another, such as from an IBM to a Macintosh.

requirements A statement of specific functions a computer program must perform, often agreed upon before the program is designed or coded.

rerun To repeat a program that has failed during execution.

reserved word A word used in an operating system program or programming language that therefore cannot be used for other purposes. For example, words like *LOAD* and *READ,* used in BASIC, cannot also be used in an application program.

reset To restart a program at the beginning, or to restart a computer.

reset button A button on a personal computer that, when pressed, starts the computer up again as if it had been turned off and on. The reset button is generally used when the computer is "locked" due to a problem in a program.

reside To be stored in.

resident font A font built into the printer hardware and thus always available for use. Fonts for dot-matrix printers are resident, and laser and ink jet printers also offer a number of resident fonts.

resident program A program that currently exists in the computer's main memory rather than in storage, and is thus available for immediate use.

resolution The clarity of a graphic image. When used in relation to printers, based on the number of dots that can be contained in an inch, with a range of 300 dots per inch to 1,200 or more. When used in relation to monitors, resolution refers to the number of pixels contained on the screen. Resolution on a standard graphics monitor ranges from 640 by 480 (640 dots on each of 480 lines) to 1,024 by 768 (1,024 dots on each of 768 lines).

resource Any aspect of a computer system, including devices like printers and modems, software, and even the technical staff.

resource allocation Designating the resources available to specific users so that, based on organizational priorities, the computer and its related resources are shared as equitably as possible.

resource sharing Making a resource, such as a printer attached to a network, available to multiple users.

response time The time required by the system to process a request from a user, measured from the moment the user presses the enter key to the moment the desired information appears on the screen. Response time varies due to factors such as the complexity of the request and the number of individuals using the system simultaneously.

restart Another term for REBOOT.

restore To return to the original form, as when a window in a graphical user interface is returned to its original size.

retrieve To obtain information from a database, performed by the computer based on a request, or query, from a user.

retrofit To modify a software application so that it fits on a specific computer system.

return In word processing, a code used at the end of a line to cause the next word of text to begin on the next line. A *soft* return occurs automatically when the end of a line is reached, while a *hard* return is inserted by a user to force text to start on the next line, as with the beginning of a new paragraph.

Return key Another name for the ENTER KEY, which sends a command entered by the user off to be processed. In a word processing program, the Return key is also used for inserting a HARD RETURN at the end of a line of text.

reusable A section of program code that can be used in other areas of the same program, or in other programs.

reverse video A screen display technology in which light characters appear on a dark background. Many standard monitors offer this option, also used with some laptop computers.

RGB monitor The abbreviation of RED-GREEN-BLUE MONITOR.

ribbon A cartridge containing a printer ribbon.

right justify To align text with the right margin of a document.

Ring Network Another term for TOKEN RING NETWORK, an approach to the design of local-area networks in which all computers in the network are connected to each other to form a ring.

RISC The acronym for REDUCED INSTRUCTION SET COMPUTER.

RJE The abbreviation of REMOTE JOB ENTRY.

robot A device that performs basic, repetitive operations like those on the assembly line in a factory; designed to respond to sensory input and can make simple, preprogrammed decisions.

robot-control language A programming language used in developing programs that control robots.

robotics The field of computer science focused on the design of robots.

robust The overall usefulness of a computer system in a variety of situations.

roll out The marketing process undertaken by a vendor in introducing a new computer or software product.

roll paper Printer paper contained on a continuous roll rather than in cut sheets.

ROM The abbreviation of READ-ONLY MEMORY.

roman A popular font style in which characters are straight rather than slanted.

ROM-BIOS BIOS that is stored on a ROM chip.

ROM cartridge A small program that performs a specific function, contained on a cartridge that is inserted into the computer's cartridge slot. Examples of programs contained on ROM cartridges include computer games.

root directory The file directory contained in the operating system, listing all of the application programs and any other files stored directly within the operating system.

rotating memory   The basic technology underlying the storage of data on hard and floppy disks, with data stored on a round device that is rotated during the process of writing and reading data.

round off   To cut off one or more digits to the right of a decimal point and increase the last remaining digit by 1 if the last digit removed is 5 or more.

route   A path through the network for sending a message from one device to another.

routine   A section of a program that performs a specific function.

routing   The process that the telecommunications system goes through in choosing a path to send a message from one device to another.

row   The horizontal line of boxes, or cells, in a spreadsheet, columns being the vertical line.

RPG   The abbreviation used for REPORT PROGRAM GENERATOR.

RS-232C   A communications standard from the Electronics Industry Association for connecting devices with serial ports.

RS-422 A communications standard from the Electronics Industry Association for connecting devices with high-speed serial ports.

ruggedized computer A computer with special design features that enable it to be used in less-protected environments, such as in military equipment.

rule Conditional instructions inserted in a program to enable it to make decisions about what action to take in specified situations.

rule line A line that runs across the screen in a word processing program, with inches marked, to help the user in setting margins.

ruler Another term for RULE LINE.

run To execute a computer program.

run in parallel To run a new version of a program together with the old version, usually as a means of testing the new version before it is used on a daily basis.

runtime The time between when a program begins executing and when it finishes.

runtime error An error that occurs while a program is being executed.

Z

SAA The abbreviation of SYSTEMS APPLICATION AR-CHITECTURE, a set of standards developed by IBM to assure that application software programs can be used across the various models and sizes of IBM computers.

SAA-compliant Software products designed to work in IBM's Systems Application Architecture.

sales forecasting Using computer-based techniques for managing the forecasting process, measuring the impact of factors such as economic climate and market growth.

SAM The acronym for SEQUENTIAL ACCESS METHOD, in which a data record is retrieved by sorting through all of the records that are stored ahead of it, one record at a time. This can be contrasted to the RANDOM ACCESS method, in which the system goes directly to the desired record.

sample data   Mock data used in testing a program or in teaching users how to use a software product.

sans-serif   Type fonts that do not include the ornate flourishes called SERIFS.

satellite computer   Another way of referring to a computer at a remote location attached to the main computer.

saturate   To fill with data, as when all the available space on a hard disk is taken.

save   To store a file on a floppy or hard disk or another storage medium.

scalable font   A font technology that produces high-quality type with an extensive flexibility in designating type sizes and shapes.

scale   To change the size of an image but not its shape, for reasons such as making it fit on a printed page.

scaling The process of changing the size of an image; in most graphics packages, a relatively simple procedure.

scan To use a device that "reads" information from a printed page and translates it into code that can be stored in a computer and manipulated like any other data.

scan area The section of a document being scanned.

scanner A device that performs the scan function.

schedule A list of planned activities in the order in which they should take place.

scheduler A software product for setting and managing schedules, based on a calendar with blocks of time the user fills with planned events. Can also refer to a software product that schedules the execution of tasks users have sent to the computer for processing.

schema The format for organizing the positions of fields on a database record.

scientific applications Software applications for the scientific and engineering communities, often requiring the ability to perform an extensive number of calculations with large numbers.

scientific notation A means of representing large numbers in a kind of shorthand as base numbers multiplied "to a power of __."

scratch To erase data.

screen The flat surface on a monitor on which the data are displayed.

screen capture Sending what is currently on the screen to a printer, generally by pressing the PRINT SCREEN KEY.

screen dump Another term for SCREEN CAPTURE.

screen flicker A situation in which the screen seems to flicker due to a condition within the monitor itself or to factors such as lighting or perceptual differences.

screen generator A software program for designing customized forms and other screen displays.

screen saver A software program that displays graphic images on the screen if the computer has been idle for more than a few minutes. This prevents BURN-IN.

script Another term for MACRO.

scroll To move line after line of data up or down the screen, or sideways, as a means of viewing the contents of, for example, a document or a spreadsheet. This process is usually accomplished with one of the arrow keys.

scroll bar A bar that appears on the side of a window in a graphical user interface. When a large file, like a document, is being displayed, the scroll bar indicates which part of the file is currently being displayed. An arrow at each end of the scroll bar is used to scroll forward or backward.

scrolling The process of moving line after line of data up, down, or sideways across the screen.

SCSI The acronym for SMALL COMPUTER SYSTEMS INTERFACE, a standard for connecting devices to the parallel interfaces on a personal computer. Used with Macintosh and some IBM compatible personal computers, it is pronounced "scuzzy."

search and replace A word processing function that changes all occurrences of a word or phrase throughout a document. The program searches for the original word or phrase and replaces it with whatever word or phrase is designated by the user.

search string The word or phrase that the word processing program searches the document for during the search and replace function.

search time The time required to search through a database and retrieve the information requested by the user.

secondary storage An additional storage area to augment a computer's main memory.

sector A section of a track on a hard or floppy disk that can hold a specific amount of data.

security Efforts undertaken to protect computer resources, including data, hardware, telecommunications lines, and software applications, from unwanted access from unauthorized users.

security administrator The individual in the organization responsible for maintaining computer system security.

security program A software application to enhance computer system security, offering, for example, a means of issuing and managing user passwords. Security programs were first developed for mainframe computers but are becoming increasingly available for personal computers and local area networks.

security software   Another term for SECURITY PROGRAM.

security utility   Another term for SECURITY PROGRAM.

seek   The mechanical process involved in locating and retrieving data from a hard or floppy disk.

seek time   The time required by a storage device to locate and retrieve data.

segment   A portion of data; used in designating storage locations for data and programs.

select   To choose an icon in a graphical user interface by using the mouse to position the arrow over the icon and clicking it. Also, the process of choosing which information to retrieve from a database.

semantics   A general term for the meanings of words and symbols in a computer program.

semiconductor   A material, made of silicon, used in building computer chips.

sense To detect a change in conditions such as current temperature.

sensing device Another term for SENSOR.

sensor Electronic devices attached to computers as input devices that can, for example, detect a change in temperature in an environmental control system and send a message to a computer that activates the air conditioning.

sequential access method (See SAM.)

sequential file organization The arrangement of files one after the other in an order based on a key data element, such as social security number or last name.

sequential processing Processing files one after the another, based on a key data element. (See also SEQUENTIAL FILE ORGANIZATION.)

serial The processing or sending of data in a sequential manner, one bit at a time, as opposed to PARALLEL, in which bits of data are sent simultaneously.

serial access A device that accesses in a serial arrangement, one bit at a time.

serial input/output The transmission and receipt of data one bit at a time.

serial interface An interface device on a computer, such as a port, in which data are sent or received one bit at a time.

serial mouse A mouse that connects to the computer through the serial port.

serial port A port on a computer through which data are sent and received one bit at a time.

serial printer A printer that can be connected to the computer through the serial port.

serial transmission Sending data one bit at a time.

serif An ornate flourish added to letters in some typefaces.

server A computer attached to a network for a single purpose. A file server, for example, contains data files accessed by other computers on the network. A print server sends print files to the printers it manages.

service Support provided to users, either by hardware and software vendors over telephone support lines, or by individuals within the organization who are trained to perform this function.

session The period of time during which the computer is in use. A session begins when a user turns the computer on and uses it, and ends when the user turns it off.

setup The overall organization of a computer system, including software applications and devices.

shade The amount of black that has been mixed in with a color.

shadowing A method of speeding up a computer's operations through a complex mix of random-access memory and read-only memory.

share To make computer resources, such as data and devices, available for more than one user.

shared logic The means by which a computer can be shared by more than one user at the same time.

shared resource A device or other resource available for use by more than one user.

shareware Software available to anyone to copy and use at will, though shareware is copyrighted and it is customary to send the software developer a fee if the software is going to be used on a regular basis.

sharpness The clarity of a computer-generated image on a screen, or of the characters produced by a printer.

sheet feeder A mechanism attached to a printer to insert sheets of paper, one at a time, for printing.

shell A user interface provided with an application program to make it easier to use—a set of menus, for example, that can be made available to users as an alternative to entering lengthy commands.

shift To change a margin in a document, or the space between columns, so that text is moved to right or left.

shift clicking An action that involves pressing the Shift key and clicking a mouse at the same time, used in some applications for selecting more than one icon at the same time.

Shift key The key on the keyboard that is pressed to capitalize lowercase letters. In application programs, the key is also used to select certain functions.

shutdown   The process of exiting the computer and turning off its power.

sift   To retrieve a small number of items from a large database.

SIG   The abbreviation of SPECIAL INTEREST GROUP, users of a specific computer or application program who meet on a regular basis and exchange information. Special interest groups can also meet online through an online information service, such as CompuServe.

sign   Another term for SYMBOL, such as those used to indicate whether a number is positive or negative.

signing on   The process of entering a computer system.

sign on   Another term for ENTERING IN, or LOGGING ON, to a computer system, which might include entering a security password.

silicon   The substance used in manufacturing computer chips.

Silicon Valley The area outside San Francisco where many high-technology companies are located.

SIMM The acronym for SINGLE IN-LINE MEMORY MODULE, which is a circuit board holding a group of memory chips, an alternative to installing one chip at a time when upgrading a computer's memory.

Simple Network Management Protocol A set of rules for information exchange between computers being used as servers and computers attached to them. (See also SERVER.)

SIMSCRIPT A programming language for creating simulated events and measuring potential outcomes.

simulation Using the computer with special application software that allows for the creation of events and conditions modeled after those that occur in real life. The computer system can then test various solutions and their impact.

simulator A special hardware device with software that simulates specific conditions, such as those used in training military personnel or pilots.

simultaneous Occurring at the same time.

simultaneous input/output The ability of a computer system to accept and output data at the same time.

simultaneous processing The ability of a computer to perform more than one operation at the same time.

single-density disk A floppy disk with a relatively low storage capacity, compared to a double-density disk or a high-density disk.

single in-line memory module (See SIMM.)

single-sided disk A floppy disk with only one side available for data storage.

sixteen-bit chip A chip that can process data sixteen bits at a time.

68000 An expression for the Motorola 68000 microprocessor, used in Apple Macintosh computers.

size The process of making a page of data, or an image, larger or smaller so that it better fits the space in which it will appear.

sketch To use a mouse or cursor to draw a series of lines, generally in a graphics application.

sketch pad A small portion of main memory, corresponding to a display area on the screen, for experimenting with images and data before they are actually saved in storage. For example, the user can hold potential graphic images in the sketch pad, choose the best one, and then send only it to storage.

skip To ignore a command or instruction.

slate PC A type of personal computer used in conjunction with an electronic pen ·rather than a keyboard or mouse. Characters are written in block style on the screen, or the slate, and the computer takes this input and translates it into standard code. Used for special applications like delivery services.

slave A device, such as a printer, under the complete control of another device.

sleeve The plastic or paper envelope in which a floppy disk is stored.

slot A rectangular opening inside a computer into which can be inserted a printed circuit board. For example, an expansion card is inserted into a slot.

small computer system interface (See scsi.)

SmallTalk A computer programming language based on designing and manipulating program objects; relatively easy to use.

smart card A tiny computer enclosed in a case about the size of a credit card.

smart system A system that makes use of artificial intelligence technology in which computers are programmed to make assessments and decisions that imitate human thinking.

smart terminal A terminal with a very limited amount of processing ability, most likely connected to a midrange or mainframe computer.

smoothing A process undertaken in a laser computer to make curves look more natural, by shifting the arrangement of dots in the curved areas.

SMS The abbreviation of SYSTEM-MANAGED STORAGE, a technology developed by IBM for storing data in very large mainframe-based information systems.

SNA The abbreviation of SYSTEMS NETWORK ARCHITECTURE, a framework developed by IBM to serve as a basis for communication between computers across a network.

SNMP   The abbreviation of SIMPLE NETWORK MANAGEMENT PROTOCOL.

soft   A general term for things that cannot be directly handled, such as *software*, in contrast to devices and equipment, or *hardware*. Items regarded as soft can also be modified more easily than those that are hard.

soft copy   Information displayed on a monitor screen rather than printed on paper, as is HARD COPY; a document displayed on the screen but not printed.

soft font   A font sent from the computer's memory to the printer hardware for printing documents. Because a soft font is not resident in the printer hardware, it is stored in the printer's memory only until the printer is turned off.

soft hyphen   A hyphen inserted by a word processing program, generally for the purpose of dividing a word at the end of a line. Hyphens specifically inserted by the user are HARD HYPHENS.

soft keys   Keys on the keyboard that, like function keys, can be used for different purposes in various application software programs.

soft return A line break performed by a word processing program as a result of the word-wrap function, which automatically moves to the next line of text when the current line is full.

soft viewing The process of looking at a document on the screen before actually sending it to the printer.

software The instructions created from computer programs that direct the computer in performing various operations. Software can also include data.

software company A company that develops and sells computer programs.

software compatibility The ability of one software program to communicate, and share data, with another.

software development The process of creating a software program, with steps that include initial design through the actual coding and on to testing and final modifications.

software documentation Printed information that accompanies a software program, with instructions for installing and using the program.

software encryption The use of software products that translate sensitive data into a code that prevents them from being read by unauthorized users.

software engineering Another term for SOFTWARE DEVELOPMENT.

software ergonomics The branch of computer science concerned with creating software programs that complement the natural abilities and reactions of human beings.

software house Another term for SOFTWARE COMPANY.

software lease A rental agreement between a software company and a user that states terms under which the user will rent the software for a specified period. Often mainframe software is leased on an annual basis rather than purchased.

software license An agreement, either formally signed or implicit, associated with the purchase of a software product, which states that the customer will use the software for his or her own personal use and will not make copies to pass on to other users.

software maintenance The "care and feeding" of a software product, including installing new product releases as well as any updates or "fixes" provided by the vendor to repair bugs in the software.

software maintenance agreement A contract between a software company and a customer describing the terms under which the company will provide telephone support, installation assistance, and updates for any "bugs" in the product. Maintenance agreements are most likely to be signed between developers and users of mainframe software.

software monitor A software product to oversee the operation of the computer system, analyze system performance, and alert the technical staff when there is a problem on the system.

software package A software program that performs a specific application, such as spreadsheets or word processing, and is offered for sale to customers who, in turn, can install it on their own systems and begin using it immediately.

software piracy Unauthorized use of a software product, generally resulting from one user making a copy of the product and passing it to another. This is a violation of copyright law and deprives the software vendor of revenue for on-going product development and support to users who have paid for these privileges.

software product Another term for SOFTWARE PACKAGE.

software protection The use of protective code in a software program to prevent it from being copied.

software publisher Another term for SOFTWARE COMPANY.

software security The use of passwords and other security measures to protect a software product from access by unauthorized users.

software support The support provided with a software product, including telephone support.

software system A general term for all the software products currently in use within an organization.

software vendor Another term for SOFTWARE COMPANY.

sort To arrange data in a specific order.

sort program A software program for arranging data in a specific order.

sound hood A plastic cover placed over a dot-matrix printer to muffle the noise it makes when printing.

source The place from which data are being taken in the process of moving them to another location. For example, when copying data, the disk from which the data are being copied.

source code Program code that has not yet been run through a compiler. (See also COMPILER.)

source disk The disk from which data are being copied to another disk.

source document A document from which information is being copied, or entered into the computer system; for example, a document being scanned by an optical scanner.

space A blank character.

spacebar The elongated key at the bottom of the keyboard, used for inserting blank spaces between words.

spaghetti code Program code that is inefficient and not very useful.

spec An abbreviated way of saying SPECIFICATION.

special character A character, such as a plus sign, that is not a letter, a number, or a symbol.

special interest group (See SIG.)

special-purpose computer A computer designed to be used only for a specific application, such as managing air traffic.

specification A requirement—for example, *design specification,* often used in defining the elements of a computer program during the initial design phase.

speech recognition The ability of a computer to recognize and respond to the human voice.

speech synthesizer A device that converts data into sounds that mimic the human voice.

spell checker A computer program that checks the spelling or words in a document. Spell checkers are either found within, or used in conjunction with, a word processing program.

spike   A sudden increase in the voltage passing through an electrical outlet.

split screen   A display screen that can be split in half for purposes such as viewing a drawing from two different angles.

spool   To wind a reel of tape.

spooler   A program that accepts files, organizes them in a line, or queue, and sends them to be processed one at a time. For example, mainframe systems use spoolers to manage files that are being sent by multiple users to a printer.

spooling   The process of sending files through a spooler for further processing.

spreadsheet   An arrangement of columns and rows that forms cells, each of which contains a value related to the other values in the arrangement.

spreadsheet application   A specific task that revolves around the use of a spreadsheet, with the columns and rows representing different variables. Spreadsheets are used for a wide range of applications, such as keeping track of monthly expenditures or calculating the effects of alternate interest rates on a loan.

spreadsheet program   A computer program for creating spreadsheets. Spreadsheet programs are primarily used on personal computers, though programs are also available for larger systems.

SQL   The abbreviation of STRUCTURED QUERY LANGUAGE, a language for retrieving data from a database that has been designed using the RELATIONAL MODEL.

SRAM   The abbreviation of STATIC RANDOM-ACCESS MEMORY, a type of RAM that is particularly fast as compared to dynamic random access memory. (See DYNAMIC.)

stack   A group of data stored in sequential order.

standalone   A computer or device that is not connected to other computers or devices and is thus not part of a network.

standard   A set of rules, or a format, applied to a specific area of technology. Standards exist, for example, to govern the establishment of networks, the storage of data, and the sending of messages over telephone lines. Standards are generally formulated by governing bodies made up of representatives from the business community, government, and education, and many of these bodies are international.

standard input The means by which a computer will receive input, generally through the keyboard, unless otherwise indicated.

standardize To design hardware or software that conforms to a set of rules established by a governing body.

standard output The vehicle through which a computer's output will be sent, generally through the display screen, unless otherwise indicated.

standby device Equipment kept maintained and readily available in the event it is needed as a replacement for equipment currently in use.

start-up The process of turning a computer's power on and going through any other required steps to make sure the computer is ready for input.

start-up disk A floppy disk containing programs, such as the operating system, to ready the computer for input.

state The current condition of a system or a device.

statement A line in a computer program that provides the computer with an instruction.

statement label A line number assigned to a statement in a computer program.

state-of-the-art The most current technology.

static Staying in one place or maintaining the current condition.

static random-access memory (See SRAM.)

station A terminal or personal computer attached to a multiple user system, such as a terminal attached to a mainframe computer.

status The current state or condition.

status report A computer-generated report indicating the current state of the system, such as use of hardware devices and software, and any reported problems. Status reports are often used by those in management positions.

storage Tapes, hard disks, floppy disks, and optical disks that can be used to hold data for future use in computer applications.

storage capacity　The amount of data, generally measured in numbers of bytes, that can be held by a specific storage medium.

storage device　A medium such as a hard or floppy disk that can be used to hold data for future use.

storage dump　To print everything currently on a storage medium as it exists, without regard to arrangement or format; usually employed to find the source of a problem.

storage protection　Security measures instituted to protect stored data from unauthorized users.

storage unit　Another term for STORAGE DEVICE.

store　To accept data from main memory and hold them for future use.

stress test　To place a software application under conditions of maximum use, with large amounts of data and a high number of users, as a means of ascertaining its reliability.

string　A line of alphanumeric characters.

string length　The number of characters in a string.

structure　Another term for ARRANGEMENT.

structure chart   Another term for FLOWCHART.

structured design   A method of developing programs based on a series of predefined, sequential steps.

structured programming   Computer programming based on principles of structured design, resulting in code that is clearly written and thus more easily modified.

Structured Query Language   (See SQL.)

stylus   A penlike device that can input data into a computer.

subdirectory   A directory one level below another directory.

subroutine   A small set of instructions within a program, which is written fro a specific function. A subroutine is *called,* or put into effect, only when a condition is met in the program. For example, in a check balancing program, a subroutine that prints an overdraft notice would only be used when a customer's account is overdrawn.

subscript   A character in a line of text that appears somewhat below the line; often used in scientific notation.

supercomputer A mainframe-sized computer that operates at a much faster speed than a standard mainframe. Supercomputers are designed to perform only a few basic applications and are used in specialized settings such as scientific research or the military.

SuperDrive A disk drive that comes with some of the models in the Macintosh line and can read floppy disks formatted both for Apple and DOS-based computers.

superscript A character in a line of text that appears slightly above the line, often used in mathematical formulas.

Super VGA A high-resolution graphics standard, with a resolution of either 800 by 600 or 1,024 by 768. Acceptance of Super VGA is rapidly increasing, and it is being supported by both personal computer manufacturers and software vendors. Super VGA is also referred to as EX-TENDED VGA.

support The assistance provided by a hardware or software company to its customers, in the form of telephone support, newsletters, and product updates.

suppress   To temporarily prevent a feature from being expressed; for example, to prevent a page number from appearing on the first page of a document.

surge   Another term for SPIKE.

surge protector   A device that protects computers and other electrical equipment from power spikes (SURGES) by providing a physical interface between the electrical outlet and the equipment. The equipment is plugged into the surge protector, which is in turn plugged into the outlet.

SVGA   (See SUPER VGA.)

switch   A button that controls the operation of a computer or other device, such as the on/off switch.

symbol   Any character that represents a function or operation.

symbol table   A list of symbols used in a software program, with the definitions of each.

synchronous   Occurring on a regular basis, as in a synchronous event.

synchronous communication A technique for rapid data transmission between computers. Because it is synchronous, each bit of data being sent is timed for transmission at a specific interval.

synchronous network A communications network designed for synchronous communication among computers, with all communications devices operating on the same time.

synchronous transmission The sending of data using synchronous communication.

syntax The rules that govern spelling and punctuation in any language, including computer languages.

syntax error An error because the user has misspelled a word or used incorrect punctuation, such as occurs when a DOS command is misspelled.

synthesizer A device attached to a computer that mimics the sound of the voice or musical instruments. Synthesizers can produce music developed on the computer through a music program.

sysop The individual who manages a computer bulletin board or a special interest group through an online information service.

system A general term for the hardware, software, data, and personnel associated with a computer-based information management system; while used in an organizational context, it can also refer to a personal computer and related hardware and software owned by an individual.

system board Another term for MOTHERBOARD.

system call A command given by the user to the operating system, as opposed to an application program.

system chart A flowchart used to document the flow of data throughout an organization's computer system.

system command Another term for SYSTEM CALL.

system flowchart Another term for SYSTEM CHART.

system folder A term used with Apple Macintosh computers to refer to the folder containing System and Finder programs.

system installation The process of setting up a computer system, including hardware and software.

system interrupt A sudden lapse occurring while
a program is being run, but that the program is
able to immediately recover and continue.

system maintanance The "care and feeding" of
the system as a whole, with a focus on assuring
that various hardware and software resources
are working together, and at peak performance,
to meet the diverse demands of the organization.

System-Managed Storage (See SMS.)

system operator (See SYSOP.)

system programmer A programmer who works
with the computer's operating system, generally
on mainframe and midrange machines. Con-
cerns of the system programmer include system
security and the reliability of data storage devices.

system programming The design of operating
systems and software products that enhance and
protect the operating system (see also SYSTEM
PROGRAMS).

system programs Computer programs designed
to enhance and protect the operating system,
installed at the operating system level. Examples

include those that manage functions such as security, and storing and retrieving data on tape, as well as programs that assist the system programmer in performing his or her job, such as report writers.

systems analyst An individual in an organization who serves as an interface between users and computer programmers, listening to the needs of users and translating these needs into a series of functions to be included in a software application program. The systems analyst works with the programmers to assure that software programs meet real-world needs.

Systems Application Architecture (See SAA.)

systems house Another term for SOFTWARE COMPANY.

systems manual Documentation describing how a hardware or software system is designed, how it is used, and what it is designed to accomplish. The term SYSTEMS MANUAL is often used interchangeably with PRODUCT DOCUMENTATION.

Systems Network Architecture (See SNA.)

Systems Resource Manager (See SRM.)

systems software   Another term for SYSTEM PRO-GRAMS.

system testing   A formal process in which mock data are loaded into a software product and all functions are used to assure that the product performs as intended under conditions simulating how it will actually be used.

system unit   Generally refers to the CPU of a personal comptuer and whatever is installed on the printed circuit boards, without the monitor, mouse, or keyboard.

T

t The abbreviation of TERA, which means one trillion.

tab To shift a line of text, generally five spaces, to begin a paragraph or create a column.

tab character A character used in documents to represent the tab, which indents the beginning of paragraphs and creates columns. The actual character used to represent a tab depends upon the word processing program.

Tab key The key on the keyboard used to perform the tab function. Generally, the tab key has the word *Tab* printed on it, with two arrows pointing in different directions.

table   A list of data, generally arranged in rows and columns.

tablet   A flat panel used with a pen input device. The user draws on the tablet with the pen, and electrical devices below the surface of the tablet's screen translate the images into input usable by the computer.

Tab Stop   The last space in a tab. Usually, a tab consists of five spaces, but most word processing programs allow the user to reset a tab so that it is narrower or wider.

Tag Image File Format   A widely used file format used for storing graphic images.

tailor-made   An application program designed to solve a specific problem, often developed by an organization's own staff of programmers.

talking computer   A computer equipped with a speech synthesizer so that it is able to verbally respond to user input.

tandem computers   Two or more computers that are wired together. If one experiences a failure, the other can take over its processing tasks, thus protecting both applications and data.

tape   A thin strip of plastic coated with a magnetic material that can store data. Tapes have large storage capacities, and, because retrieving data from a tape is a relatively slow process compared to doing so from other storage media, they are most often used for storing data that are no longer being used on a regular basis.

tape cartridge   A device shaped like a large cassette, containing a magnetic storage tape and used with midrange and mainframe computers.

tape cassette   A device similar in size to a small cassette, containing a magnetic storage tape and used with personal computers.

tape drive   A device that reads data from, and writes data on, magnetic tape.

tape label   Information contained on a magnetic storage tape identifying the data contained on the tape and the date the data were stored.

tape library   A physical storage area, often located near the computer, where magnetic storage tapes are labeled and stored for future use.

target disk   The disk on which data are copied during the process of copying data from one disk to another. The disk the data are copied from is the SOURCE DISK.

task   A job being performed by a computer, generally as the result of a command entered by the user.

TB   The abbreviation of TERABYTE, which is one trillion bytes.

TCP/IP   The abbreviation of TRANSPORT CONTROL PROTOCOL/INTERNET PROTOCOL, a widely used set of rules for communications between computers across a network. TCP/IP was developed by the Department of Defense.

tear-off menu   A menu that appears as a result of choosing an option from a menu bar; similar to a pop-up menu but can also be moved to a different area of the screen.

technical support   Another term for SUPPORT, referring most specifically to the telephone information lines manned by product specialists that hardware and software vendors make available to their users.

telecommunications   A general term for the transmission of any kind of data, including voice and images, over a telephone line.

telecommute The practice whereby employees work in their own homes and perform their job tasks by accessing their organization's computer system through a personal computer and modem.

teleconference A meeting that takes place among individuals in different geographical regions, all connected through telephone lines carrying voice and, increasingly, video images.

telecopy Another term for sending a facsimile message.

Telenet A large public data network owned by U.S. Sprint Communications.

telephony A general term for the underlying technology of the telephone, including the conversion of sound into signals that are transmitted to other locations and then converted back into sound.

teleprocessing Transmitting data from one computer to another through the use of telecommunications technology.

teletext The information, including images, sent from a television station to the television set of each individual viewer.

teletypewriter A telecommunications device consisting of a typewriter keyboard with a built-in printer and modem. It is connected to a computer through a telecommunication line and can both receive and sent data. Teletypewriters have generally been replaced by personal computers with modems.

template A pattern or guide; can be used to refer to the plastic guide used in creating flowcharts. In graphic software programs, sample images that can be used as is or modified. In spreadsheet programs, sample spreadsheets that the user can complete with his or her own data.

temporary file A file in main memory used to store information during a session, but that disappears when the user leaves the program or turns the computer off.

temporary storage The area of main memory in which temporary files are stored. (See also TEMPORARY FILE.)

terminal A display screen and keyboard attached to each other with no processing power of their own, unlike a personal computer. Terminals are usually attached to a mainframe computer.

terminal emulation Making one terminal act like another. Often, the process of connecting a personal computer to a mainframe computer and then using the personal computer as if it were a standard mainframe-attached terminal.

terminal session Another term for SESSION, generally when a user employs a terminal attached to a mainframe.

test data A set of sample data used with a program to assure that the program operates according to specification.

testing The process of observing how a software program, or hardware device, operates under conditions that approximate the environment in which it will normally be used.

text Data that consist of letters and numbers arranged in words and sentences. Text is usually associated with information arranged in documents.

text editor A software program used in editing and manipulating text data. Text editors can perform basic operations, such as formatting, but are not as full-featured as a word processing program.

text file A file containing data arranged as text, such as a word processing file.

text mode   Used when referring to a monitor that will only display characters, rather than graphic images.

text processing   Using a computer to retrieve and manipulate text files such as documents.

texture   Shading added to a graphic image that gives it the appearance of being multidimensional.

thermal printer   A printer that produces an image on heat-sensitive paper through heated pins pushed against the paper; used in facsimile machines.

thesaurus   A computer program containing words and their synonyms, and used in conjunction with a word processing program.

third party   A company that develops hardware or software products that complement those of another company. For example, a software company that develops applications for use with IBM computers would be described as third party.

three-dimensional graphics   Graphic images displayed on the screen in three dimensions, including height, width, and depth. Producing three-dimensional graphics requires the use of graphics software and a display screen with this capability.

three-dimensional spreadsheet A spreadsheet program that allows the user to display up to three separate spreadsheets, making it easier to make comparisons among different sets of information.

386 The abbreviation of INTEL 80386 MICROPROCESSOR.

386SX The abbreviation of INTEL 80386SX MICROPROCESSOR.

throughput The amount of processing that a computer can perform during a specific period, the rate of which often serves as a basis for comparisons among mainframe systems.

thumbnail A display of multiple text pages on one screen, in desktop publishing programs, used to check overall format of each page before it is sent off to a printer.

TIFF The acronym for TAG IMAGE FILE FORMAT.

tiled windows Windows arranged on the screen as adjacent blocks with no window overlapping another.

time-sharing   The use of one computer, generally a mainframe, by more than one simultaneous user. In a standard time-sharing system, a service bureau company offers the use of its mainframe computer to other companies, who access it for a fee. This was once a standard arrangement, but as the price of computing power declined during the 1980s, more companies could afford their own mainframe computers, and the time-sharing industry declined.

title bar   The bar across the top of the screen in many graphical user interfaces, containing the name of the application.

toggle   A two-stage switch that can be turned back and forth from one setting to another. For example, the Caps Lock key, when pressed, causes all alphabetic characters to appear in uppercase until it is pressed again; subsequently typed alphabetic characters appear in lowercase.

token   In a token ring network, refers to the message that is passed through the communications line.

token passing   If a message is being sent across a token ring network, it is attached to the token with the address of the computer to which it is being sent. The token passes from one computer to the next, delivering a message to one computer, and picking one up from the next.

token-ring network   A local-area network in which personal computers are arranged in a circle, or ring, with messages, or tokens, passed through the ring.

tone   Another term for the SHADE of a color.

toner   The dry, powdery ink used in laser printers.

tool   A software application product.

toolbox   A software package consisting of a set of programs that can be used by computer programmers in developing applications; often includes time-saving features such as standard routines that can be inserted into a program to shorten development time.

toolkit   Another term for TOOLBOX.

top-down design   A computer program design methodology that begins with a detailed description of what the program should accomplish functionally. These functions are then broken down into tasks. Each separate task is assigned to a programmer, or a group of programmers, to translate into program code.

top-down programming   The actual process of developing a program that results from a top-down design. The programmer turns each task into series of routines and then subroutines, working from the top down.

topology   The actual physical shape of a local-area network. TOKEN RING is an example of a topology.

touch-sensitive screen   A display screen on which the user can make selections by touching the screen, which has an electrically charged surface sensitive to touch. Often used in retail settings, such as maps in shopping malls.

touch screen   Another term for TOUCH-SENSITIVE SCREEN.

tower configuration   A personal computer in which the actual CPU, including the motherboard, expansion boards, and disk drives, is mounted in an upright cabinet rather than the standard desktop cabinet. Towers can be placed on the floor, beside or under a desk, and thus conserve space on the desktop.

TPI   The abbreviation of TRACKS PER INCH, the number of tracks on a magnetic disk. The more tracks, the more data that can be stored. High-density floppy disks have ninety-six tracks.

trace   To make an ongoing record of the operation of a program, with each action resulting in a statement that describes what was done; determines the exact location of program bugs.

track   A circular area on a disk in which data can be stored; made in a disk as a result of the format process.

trackball   A pointing device consisting of a small platform with a ball resting on it, similar in size to a mouse. The platform remains stationary, while the user manipulates the ball with his or her hand, and thus moves the cursor or arrow on the screen.

tracking   The process of using a mouse, trackball, or other pointer device to move a cursor or arrow around the display screen.

tracking symbol   The symbol on the display screen—generally an arrow—indicating the current position of the cursor.

tracks per inch   (See TPI.)

tractor feed   A device for passing a continuous roll of paper through a printer. A tractor feed has small pins mounted on wheels, one at each end of the printer, that catch in the small holes lining each side of the paper and pull it through.

traffic   The number of messages being carried over a communications line.

transaction   A unit of activity that takes place in a computer system, such as when a user enters a command to retrieve information from a database.

transaction processing   A method of processing user transactions in which, once a request is made, the computer system responds immediately, rather than at a later time as in batch processing. For example, if a user makes a change in a database record, the record is updated immediately rather than at a later time.

transceiver   A device that can both receive and transmit data, and that connects a computer to a local area network.

transcribe   To copy data from one location to another, often requiring that the data be converted to a new form, such as from handwriting to code understood by the computer.

transfer To move data from one location to another.

transfer rate The time required to move data from one location to another; often used when measuring speed on a network.

transfer time The amount of time occurring between when a message is sent and when it is received.

transient The area of main memory that contains whatever program is currently being used, only to be replaced by the next program the user chooses.

translate To convert a program written in a high-level language, like COBOL, into a machine language, like Assembler, through the use of a compiler. Another term is COMPILE.

translator Another term for COMPILER.

transmission The sending of data over a communications line from one location to another.

Transmission Control Protocol/Internet Protocol (See TCP/IP.)

transparent Processes performed by the computer system that are relatively invisible to the user. Generally, an operation performed by a software application program, based on a simple command from the user, is transparent.

transportable computer A computer that can be moved among different locations; also referred to as a PORTABLE COMPUTER.

troubleshoot To look for the cause of a problem in a hardware device or a software program; similar to DEBUG.

truncate To cut off the final characters in a word, or the final digits in a number.

trunk A direct channel between two telephone switching stations.

TTY The abbreviation of TELETYPEWRITER.

turnaround time The time required for a user's request to be processed by the computer system and an answer returned.

turnkey system A complete computer system with all necessary hardware and software, documentation, and even training, delivered to a user as a ready-to-use solution. All the user has to do is "turn the key" and get started.

turn off   To press a computer's off switch and thus cut off its electrical or battery power.

turn on   To press a computer's on switch and thus provide it with power.

tutorial   A teaching device provided with a hardware or software product. The tutorial may be in the form of a manual or on a disk that includes exercises simulating the actual use of the product.

tweak   To make final adjustments in a software program, or to adjust a hardware device to raise its level of performance.

twisted-pair cable   A cable consisting of two wires bound together and used for data transmission. One wire carries the data signal while the other absorbs other signals that would interfere with data transmission.

twisted wire   Another term for TWISTED PAIR CABLE.

286   The abbreviation of INTEL 80286 PROCESSOR.

Tymnet A large public data network owned by McDonnel Douglas.

type To input data into a computer through a keyboard. Type can also refer to a kind of data, such as alphanumeric characters.

typeface A style of type, such as Times Roman or Courier.

type font In printing, a set of characters with a unique appearance based on factors such as typeface, type size, type style, and boldness. Typeface is one aspect of the type font.

typeover Another term for OVERSTRIKE.

type size The size of typed characters, usually measured in points.

type style Generally, how much a typeface slants to the right, as in whether or not the typeface is in italics or not. Type style can also refer to TYPEFACE.

U

UART The acronym for UNIVERSAL ASYNCHRONOUS RECEIVER-TRANSMITTER, a single device that integrates the circuits required for both receiving and transmitting communications in an asynchronous mode. Commonly used for connectivity between personal computers, as well as for a personal computer and a modem.

ultralight computer A small portable computer, approximately the size of a notebook and weighing 4–6 pounds.

unattended operations The ability to operate a computer and related devices without a personnel staff; generally used in relation to midrange and mainframe computers at remote locations.

unbundled A software product sold as a separate entity rather than as part of a larger system with hardware and other software products.

undelete A command to restore information that has been deleted in the previous delete command.

underflow The condition resulting from an attempt by the computer to produce a number that is too small, that is, too close to zero. Generally, this results in an error message.

underline A word processing function to underline words or blocks of text.

undo A command to reverse the the action resulting from the previous command. Undo is similar in result to undelete, except that undo is not limited to restoring deleted information.

undocumented Software programs, or, more likely, features within programs, that are not described in product manuals or other documentation. Often these features are not intended for use on a regular basis.

unerase Another term for UNDELETE.

uninterruptible power supply A battery-operated power source attached to a computer and designed to take over in the event of an electrical power failure.

unit A single item. A general term for a hardware device or a piece of data.

universal asynchronous receiver-transmitter (See UART.)

universal language A computer programming language such as BASIC that can be used on computers developed by different hardware manufacturers.

universal product code A code that can be read by a computer, used in labeling retail products.

UNIX An easy-to-use operating system able to accommodate the needs of many users at the same time. Originally developed by AT&T, can run on all sizes of computers, from personal computers to mainframes.

unmount To remove a tape from a tape drive, or a floppy disk from a disk drive.

unpack To convert packed data to its original format. (See also PACK.)

up A slang term for a computer system that is turned on or otherwise ready for active use.

update To install a new release of a software program, so that the latest features are available for use.

upgrade Another term for UPDATE.

upload To transfer data from the main computer to a computer at a remote site, as when a file is being sent from a main office to a user at home.

uppercase Another term for CAPITAL LETTER.

UPS The abbreviation of UNINTERRUPTIBLE POWER SUPPLY.

uptime The time during which a computer system is available for use.

upward compatible A software product that is designed so that it will still be usable on a future system with a more powerful operating system and hardware; for example, a word processing program that can run under both the current version of DOS as well as new releases.

USENET An international network for information sharing among UNIX users.

user   Anyone who uses a computer, regardless of his or her level of technical expertise, though the term is usually associated with limited computer knowledge.

user-defined   Features in an application program that can be changed or modified by the user. The term can apply, for example, to a program's print function, in which the user decides which printer and font style will be used.

user-defined function key   Another term for PRO-GRAMMABLE FUNCTION KEY.

user-friendly   A function of a software program or hardware device that is easy for individuals with limited computer experience to use.

user group   A local or national association of computer users with similar professional or hobby interests, or who use a specific hardware or software product. Generally, user groups meet periodically to exchange information and also publish newsletters.

user interface   The part of a computer program that the user sees and makes selections from, as opposed to those parts executed internally in the computer. Usually composed of graphic screens with icons, or of a series of menus from which the user makes choices.

user manual   Documentation accompanying a software or hardware product that is written in basic, nontechnical language aimed at the user. The user manual often describes the major functions of the product accompanied by step-by-step instructions.

user memory   The area of a computer's main memory available for application programs and data. This area is separate from the area reserved for the operating system.

username   The name by which a user identifies himself or herself in a network. The user may be able to choose a username, or it may be based on his or her initials, or it may be assigned.

user profile   A basic list of information about each user in a computer system, usually maintained by in individual in the MIS department. The user profile includes information such as username, the user's physical location, and the applications and databases that the user is authorized to access.

utility   A small program that performs a very specific function for a user, such as assisting a programmer to edit and debug a computer program.

utility program A program that performs a specific function at the system level, such as managing the use of magnetic storage tapes for system backup.

V

VAB The abbreviation of VOICE-ANSWER BACK, the use of voice recordings by the computer to respond to requests by users; often used by banks for telephone customer service lines.

validation The process undertaken in a database management system to ensure that any data being entered in a database have the correct format. For example, a name field can most likely contain only alphabetic characters and not numbers.

validity checking The same as VALIDATION.

value-added network A network that offers, not only a means of transmitting data, but also additional services such as electronic mail.

value-added reseller A company that sells complete systems, with both hardware and software, and also offers additional support and installation services.

VAN The abbreviation of VALUE-ADDED NETWORK.

vaporware A slang term for software that has been announced to the market, such as through advertisements, but is still being developed and thus not actually available for purchase.

VAR The abbreviation of VALUE-ADDED RESELLER.

variable-length field A field in a database record that can vary in length based on the data stored in it.

variable-length record A record in a database that can vary in length based on the data stored in it.

VDT The abbreviation of VIDEO DISPLAY TERMINAL, a device consisting of a screen and a keyboard, used to interact with a computer.

VDT radiation A form of radiation emitted by video display terminals. It is not believed to be dangerous.

version A specific release of a software product. Each new version, or release, of a software product offers the basic product capabilities with additional features.

version number Each new version, or release, of a software product is assigned a number, and the versions are numbered consecutively, on an ascending order, generally beginning with the number 1.0. Examples of DOS versions include DOS 3.0 and DOS 5.0.

vertical scrolling Moving a document up and down on the display screen to view different sections.

VESA An association of companies who manufacture computer monitors and related video equipment, who meet periodically to set industry-wide standards such as VGA.

VGA The abbreviation of VIDEO GRAPHICS ARRAY, a video adapter standard for IBM and compatible computers with graphics resolution of 640 by 480 with 16 colors, and 320 by 200 with 256 colors. VGA is a more recent standard than CGA and EGA.

VGA Plus Another term for SUPER VGA.

video adapter   Electronic components mounted on an expansion board and inserted into a slot in the computer to provide it with the ability to display a video image.

video board   Another term for VIDEO ADAPTER.

video controller   Another term for VIDEO ADAPTER.

videodisc   An optical disc that stores a video image accompanied by sound.

video display   A monitor or terminal capable of displaying graphic images.

video display board   Another term for VIDEO ADAPTER.

video display unit   A display monitor capable of displaying video images.

Video Electronics Standard Association   (See VESA.)

video game   An interactive game, like a computer game, played on a video screen or television with either a joystick, mouse, or keyboard.

video graphics array   (See VGA.)

video memory   An area of memory in a video system that stores the actual video image.

video mode   The mode, or state, in which a video adapter can be set. The video adapter will generally display an image in either graphic or text mode. In text mode, alphanumeric characters and symbols can be displayed, but not images. In graphics mode, images can also be displayed.

video signal   A signal sent from a video adapter to the display screen, controlling the shape and brightness of the video image.

video standards   Industry guidelines for the development of video adapters, including EGA, CGA, and VGA.

video terminal   A terminal designed to display video images, generally attached to a midrange or mainframe computer.

view   To display information on a video screen, such as a document or a graphic image. Different angles of the same graphic image are referred to as different views.

virtual   A general term in computer science for anything that is not tangible or real.

virtual disk Another term for RAM DISK.

Virtual Machine/System Product An operating system designed by IBM for use on mainframe computers, characterized both by ease of use and the ability to accommodate multiple users simultaneously.

virus A program that, once it enters a computer system, can cause damage either "infecting" other files with bad code, or rapidly replicating itself so that all available memory is used and the system is brought to a stop.

visible page The graphic image currently being displayed on the monitor screen. Each separate graphic image is referred to as a *page* for storage purposes.

VM/SP The abbreviation of VIRTUAL MACHINE/SYSTEM PRODUCT.

voice answer back (See VAB.)

voice input The use of the human voice as input into the computer.

voice mail A telephone-based electronic message system, with telephone messages converted into a format that can be stored in, and retrieved from, the computer system.

voice output   The ability of the computer to respond to users with words that can be understood by the human ear, spoken through a speech synthesizer.

voice recognition   The ability of the computer to recognize and respond to human speech. This ability is generally limited to accepting basic commands in a carefully spoken voice, rather than actually conversing.

voice synthesizer   Another term for SPEECH SYNTHESIZER.

volatile memory   Memory, such as RAM, that loses its contents when the computer is turned off.

volatile storage   Another term for VOLATILE MEMORY.

voltage surge protector   Another term for SURGE PROTECTOR.

volume   A physical device for data storage, such as a floppy disk or a magnetic tape.

volume label   An identifying name provided by the user, given to a floppy disk or a magnetic tape generally at the time it is formatted.

volume name   Another term for VOLUME LABEL.

volume number   Another term for VOLUME LABEL.

volume reference number   An identifying number assigned to a disk or tape by the system, generally at the time it is formatted.

VRAM   The abbreviation for VIDEO RAM.

W

wafer A circular-shaped semiconductor disk, about three inches thick, used in creating integrated circuits. Multiple circuits are fabricated on the surface of this crystal material, and then it is cut up into individual chips.

wait state A period of time, often so short as to be unnoticeable, in which the CPU is idle, during the execution of an instruction, while it waits for data from memory or storage. This happens because the CPU itself may be faster than the input devices.

WAN The abbreviation of WIDE-AREA NETWORK, a network of computers spread out over a large geographical area, across states and even countries.

wand   A pen-shaped device used to scan bar codes as input into a computer.

warm boot   To restart the computer system without actually turning the computer off and on, by pressing either the restart button or the Alt, Control, and Delete keys simultaneously. Generally, a warm boot begins with loading the operating system.

warm start   Another term for WARM BOOT.

watt   The standard unit of measurement of electrical power.

"what-if" analysis   The most common type of spreadsheet analysis, in which one or more values are changed and the results calculated to better understand the future effect of these changes. For example, the effects of various interest rates can be assessed by recalculating the spreadsheet values based on each rate.

what you see is what you get   A word processing term describing the ability to view a document on the display screen and have it appear almost exactly as it will appear on the printed page, with all page numbers, spacing, and other formatting. The acronym is *wysiwig,* pronounced "wizzywig."

wheel printer Another term for DAISY-WHEEL PRINTER.

whitespace Areas on a display screen, or on a printed page, that are free of either type or images.

wide-area network (See WAN.)

wideband Another term for BROADBAND.

widow In word processing, the last line of a paragraph appearing at the top of the next page. Widows can be avoided by adding a page break earlier in the paragraph.

wild card character A character such as * (asterisk) that, in some programs, can be used to substitute for one or many other characters. For example, often used in searching documents or files with "b*", causing the program to retrieve a word that begins with a "b," regardless of the number of characters in the word.

Winchester disk drive A hard disk drive developed by IBM for use in personal computers.

window   A section of the screen, in a graphical user interface, containing icons or a list of options that can be selected by the user with a mouse.

windowing environment   A graphical user interface based on windows.

Windows   Generally, Microsoft Windows, a graphical user interface with multitasking capabilities for use with Microsoft's MS-DOS operating system.

word   Any group of characters that forms a unit of meaning and is separated from other groups of characters.

word processing   The process of creating, editing, and storing documents with a computer equipped with word processing software.

word processing program   Application software designed to perform word processing, with features that allow the user to type a document, edit the text, format it with page numbers and appropriate spacing, print it, and store it for future use.

word processor   Another term for WORD PROCESSING PROGRAM.

word wrap   A standard feature in a word processing program that enables the user to easily keep text within the specified margins. With word wrap, when the user is typing a word at the end of a line that will go over the margin, the program automatically moves the word to the next line. (See SOFT RETURN.)

workgroup computing software   A software application program that enables individuals working on a specific task, like the members of a department, to share information and scheduling activities. Workgroup computing software generally offers a range of functions including electronic mail, a calendar, and the document-editing capabilities that allow multiple users to comment on the same document.

working directory   The directory of files the user is currently involved in using—for example, in using a word processing program, the directory of files associated with that program.

worksheet   Another term for SPREADSHEET.

workstation   Broadly, applies to any computer available for use by only one individual at a time, and as such can refer to a personal computer; however, generally assumed to refer to high-power, full-featured desktop computers used for scientific and engineering applications. These are often based on the UNIX operating system with high-resolution screens, fast processing power, and large storage capacities.

WORM The acronym for WRITE-ONCE, READ-MANY, an optical disk on which data can be written only once and not subsequently changed or erased. Used for storing large amounts of data, and related uses in which data will not change once they have been recorded.

WP The abbreviation of WORD PROCESSING.

write To transfer data from main memory to a storage device such as a floppy disk or magnetic tape. Also, the process of transferring data from main memory to the computer's display screen.

write error An error that occurs during the process of transferring data from main memory to the display screen or a storage device.

write-once, read-many (See WORM.)

write-protect To physically protect a floppy disk or tape to prevent data from being recorded on it. For example, a floppy disk containing an application program is write-protected to prevent the program from being accidentally destroyed.

write-protect notch   A small indented space in the jacket of a floppy disk which, when covered with a piece of tape, prevents data from being written on the disk while still allowing data to be read from it.

write-protect ring   A plastic ring used with a magnetic tape to prevent data from being written on the tape.

write-protect tab   Another term for WRITE-PROTECT NOTCH.

WYSIWYG   The acronym for WHAT YOU SEE IS WHAT YOU GET.

X

X.25 A communications standard from the CCITT (Consultative Committee in International Telephone and Telegraph), an international communications standards organization. Developed to guide the establishment of local area networks using PACKET-SWITCHING as a basis for sending and receiving data.

X.400 A communications standard from the CCITT (Consultative Committee in International Telephone and Telegraph) providing protocols for sending electronic-mail messages across an open network. (See also ELECTRONIC MAIL.)

X.500 A communications standard from the CCITT (Consultative Committee in International Telephone and Telegraph) that extends X.400 through guidelines providing directory services across an open network. Through these directory services, different electronic-mail systems can be connected together so that messages can be sent from one electronic-mail system to another.

Xenix   A form of the UNIX operating system for use on personal computers, originally developed by Microsoft.

XGA   The abbreviation of EXTENDED GRAPHICS ARRAY, a graphics standard that supports a higher resolution than VGA.

Xmodem   A protocol used in transferring data over telecommunications line through the use of a modem. It is recognized as being relatively easy to use as well as reliable, sends data in blocks, and checks to make sure a block of data has been sent and received before sending another.

XMS   The abbreviation of EXTENDED MEMORY SPECIFICATION, a method by which the standard amount of main memory under DOS can be extended to enhance an application program's performance. XMS was developed by Lotus, AST research, Microsoft, and Intel.

X Windows A system for developing graphical user interfaces with windows, used primarily with UNIX-based systems; developed at the Massachusetts Institute of Technology.

Y

Ymodem A protocol used in sending data over a telecommunications line, very similar to Xmodem, but with technical enhancements including the ability to send a group of files at the same time.

Z

zap To make an error that destroys the contents of a file.

zero The numeral with no magnitude.

zip A standard DATA COMPRESSION format.

Zmodem A protocol for sending data over telecommunications lines through the use of a modem. Zmodem offers enhancements over Xmodem including the ability to send larger amounts of data at a faster rate.

zoom With graphic user interfaces, the process of making a window larger, so that it fills more of the screen.

Index of Computer Terms

Each area of high technology has its own separate vocabulary. This index should help users clarify what that specific vocabulary is. The categories allow users to explore groups of words employed in areas of special interest, and to locate the particular words they interface with in particular applications.

Database and Spreadsheet

batch
browse
cell
chain
chaining
change file
close
column
data attribute
data bank
database
database administrator
database management system
tem
database manager
database server
data capture
data dictionary
data element
data entry

data manipulation
data structure
DB
dbms
distributed database
field
file
file attribute
grid
gridsheet
HDBMS
header record
hierarchical
index
inquiry
join
keyword
list
list processing
matrix

Index

Graphics and Video

Index

Large Systems

Including mainframe and midrange computers, and concepts related to computing in a large organization

Index

Networking and Communications

At any hardware level, from local area networks to multiple mainframes

Index

Operating Systems

Including DOS and UNIX

Personal Computers

Including hardware specifically related to personal computers, as well as laptop, notebook, and other portable computers

Index

Printers and Peripheral Devices

Including fax machines and modems, and the hardware and software terms related to connecting and using these peripheral devices

Index

sensor
serial printer
sheet feeder
slave
smoothing
sound hood

standby device
thermal printer
toner
tractor feed
wheel printer

Programming

Including any terms related to computer programming/software application development

action statement
Ada
AND operator
annotation
annotation symbol
ANSI.SYS
API
application program
application program interface
application programmer
application programming
argument
arithmetic expression
arithmetic operator
artificial language
assemble
assembler
assign
automatic programming
back end
backward compatible
base address
BASIC
batch program

binary
binary code
binary file
Binary Format
bit
Boolean expression
Boolean logic
Boolean operator
bug
byte
C
C++
call
case sensitivity
COBOL
code
code conversion
coder
code segment
coding
coding form
comments
compile
computer-independent language

Index

Word Processing and Desktop Publishing

Index

THE PRINCETON LANGUAGE INSTITUTE is a consortium of experts comprised of linguists, lexicographers, writers, teachers, and businesspeople. Applying academic rigor to practical endeavor, the Institute enables writers and members of professional communities to enhance their communication and language skills as they work to meet the challenges and complexities of the 21st century. The Princeton Language Institute is based in Princeton, New Jersey.

GARY McCLAIN, PH.D. is an expert in the field of technological information. Since 1981 he has taught the use of software products, designed applications of software systems, and conducted multilevel market research for high technology clients. He presently holds the position of Vice President of Technology at International Techvantage Group, a New York City-based market research firm.

He is the author of six books on computer technology and information research, including the *Henry Holt International Desk Reference,* as well as numerous articles on similar subjects, and is co-author of *State-of-the-Art Fact-Finding*.